Fate & Philosophy

Fate & Philosophy

A JOURNEY THROUGH LIFE'S GREAT QUESTIONS

JIM FLYNN

AWA PRESS

First edition published in 2012 by Awa Press, Level One,
85 Victoria Street, Wellington, New Zealand.

ISBN: 978-1-877551-32-1

ebook formats
epub 978-1-877551-43-7
mobi 978-1-877551-44-4

National Library of New Zealand Cataloguing-in-Publication Data
Flynn, James Robert, 1934-
Fate and philosophy : a journey through life's great questions / Jim Flynn.
Includes index.
1. Philosophy. 2. Ethics. 3. Science—Popular works.
I. Title.
100—dc 23

Front cover photograph by Felbert Eickenberg, Getty Images
Back cover photograph © franckreporter, iStockphoto
Book design by Keely O'Shannessy
Typesetting by Tina Delceg
Printed by Everbest Printing Company Ltd, China
This book is typeset in Minion

www.awapress.com

Published with the support of

ARTS COUNCIL OF NEW ZEALAND TOI AOTEAROA

To Alan Musgrave

*friend and colleague
and mad dog realist*

Man is what he believes.

ANTON CHEKHOV
NOTEBOOK ENTRY

It matters not how strait the gate,
How charged with punishments the scroll,
I am the master of my fate;
I am the captain of my soul.

WILLIAM ERNEST HENLEY
"INVICTUS", 1875

Men are not prisoners of fate,
but only prisoners of their own minds.

FRANKLIN D. ROOSEVELT
PAN AMERICAN DAY ADDRESS, 1939

CONTENTS

ON BECOMING A PHILOSOPHER

Every year students enter my courses with a collection of attitudes and opinions. The most common ones are that religion is silly and boring; people should be tolerant of one another's values and not be judgmental; what is natural is best; "I wonder if the scientists know what they are doing?"; and of course no one has free will.

Some of these students leave as altered beings, interested in the mystical experience (for some a religious experience), committed to humane ideals and knowledgeable about how to defend them, accepting that no one worships nature without reservations, cognizant of the foundations and achievements of science, and aware of just how important it is to think through the issue of free will vs. determinism.

Unless you learn to reason about what you believe you are a prisoner of fate. "Your" views are no more than the views of society or your parents or perhaps your church. It is sterile to simply become a cynic and rebel against these people and institutions. That kind of rebellion usually means no more than that your generation has different opinions from those of the older generation.

This book invites you to move on, look within, and discover the person you really are. Do you really need someone else or something, perhaps a god, bible or nature, to tell you what is good?

Philosophical knowledge is much more fragile than other kinds of knowledge. It is unlikely we will cease to know how to light fires or forget what we know about evolution—at least without a descent into mindless barbarism, as with Stalin and born-again Christians—but love of money blinds people to what sort of life is best for human beings. Similarly, the very potency of science may polarize us into those who trust science without reflection and those who fear and loathe it. A decline in genuine religious experience may beget both village atheists and churches full of nonsense.

If sound philosophy is so much at the mercy of history, how likely is it that you have been born at the one time and place over the last two thousand years when conventional opinions tell the uncontaminated truth? You would have a better chance of winning a national lottery.

If you are free of the tyranny of desperate poverty, spies of the totalitarian state, chronic unhappiness and mental illness, the most important liberty you can enjoy is freedom of the mind. Everyone who reads this has a philosophy: everyone makes assumptions about what exists, what is known, whether it makes sense to blame people, and what they ought to do. You have two options: you can accept uncritically the philosophy you have, or you can think for yourself about certain basic questions. These questions determine how you live out your life. As Anton Chekhov said, "Man is what he believes."

I was fortunate enough to become reflective about what I believed just in time. I was born into the Catholic Church and as a young child believed everything without question. When I was ten I began to read an encyclopedia called *The Book of Knowledge*. It told me that the sun and planets had coalesced out of a nebula of gas, and human beings had evolved just like all other animals. Two years later, when Father Burns, my confessor, asked me whether I would like to go to St Charles in Baltimore, a high school whose pupils intended to become priests, I had enough doubts to say no. I knew that once at St Charles it was not impossible to leave but it was traumatic: you had to either be dishonest or tell everyone you thought they were dedicating their lives to a myth.

You may say the way in which becoming reflective influenced my life was unusually dramatic, but the great satisfaction of questioning what you believe is the sensation that you have some control over your fate. There is a deep sense that you are beginning to find yourself—the wiser self that lay hidden behind the unreflecting self that was sleepwalking through life. This life-changing experience is available to everyone.

How much philosophy will change your opinions depends on what kind of sleepwalker you are. Understanding ethics may cure you of wanting some authority to tell you what is right and wrong. Understanding science will inoculate you against a whole legion of nonsense, from astrology to the Bermuda Triangle to whether people can use psychic powers to bend spoons. Understanding religious experience will inoculate you against childish concepts of God.

Perhaps you already respect science, are happy to judge right and wrong for yourself, and are an atheist. Very well, but how do you know it is reasonable to be that kind of person until you can justify your opinions in the light of reason?

By the end you may decide it is too disturbing to think about issues so intensely and prefer a quiet life. Then I may have done you a disservice: it is easier to open the Pandora's Box of philosophy than to shut it. The problems discussed here may become enemies of sleep, things you cannot get out of your mind without thinking them through. However, you will have been transformed into a human being who has a right to believe that certain things are true because you can articulate why they are true.

The purpose of this book dictated the style. Although I hope my professional colleagues will find it interesting, it is written for people who have not yet begun the systematic study of philosophy, or are near the beginning of their investigations. I do not defend myself against all possible objections to my conclusions but the reasons I give are my real reasons for holding the views I do. I hope this book will be particularly helpful to students who have taken one year of philosophy and are deciding whether to continue.

The book's purpose also set the content. For example, I have not included a chapter on aesthetics, or the theory of beauty. You can get through life without this. You are ravished by Mozart's operas but cannot wait for *Doctor Atomic* to end. You can stand for half an hour in front of an

El Greco, but would rather be home looking at the wallpaper than at "White on White". However, you cannot get through life without living as if God either does or does not exist, or wondering what sense it makes to be good, or praising and blaming other people, or wondering if a good society is possible.

This book discusses problems in what I see as a logical order. The first, second and third chapters address the age-old human yearning to find someone or something to tell us what is good. The fourth and fifth seek to show that the alternative is not so bleak: even without such an authority we need not conclude that no moral ideals are any more defensible that any others. The sixth shows that, despite arguments to the contrary, a good society is possible, and the seventh that free will is both coherent and may exist. The eighth chapter looks at whether it is appropriate to praise or blame people for their behavior.

The ninth chapter examines the most fundamental objection raised against science, namely the problem of induction. If spontaneous events can occur, is it not a matter of faith whether we can trust science to predict the future? This problem cannot be solved, but I try to show it is a special case of a larger problem that afflicts all theories of being. It will plague you no matter whether you believe in God or think that only the physical universe exists. Therefore, its non-solution lends no school of thought an advantage. It simply has to be set aside if we are to develop any metaphysics or theory of existence whatsoever.

The tenth, eleventh and twelfth chapters endorse realism. I believe that everyday life and science show things existed prior to and beyond human experience, even though they register their existence to us only through human experience, supplemented by the instruments we have invented. I also argue that the physical universe—and perhaps alternative physical universes—exist in a way that is unique. Other things, such as the truths of mathematics, "exist" only in a special sense: they exist in our minds and cannot affect the physical world without the help of a human agent.

The thirteenth chapter updates the arguments for the existence of God and shows why they are defective. The fourteenth analyzes the mystical experience, an unusual kind of experience in which a person apprehends something not present to the five senses. This kind of experience has the best case to qualify as a knowledge-giver about a thing that exists in addition to the physical universe or is an attribute of the physical universe we normally miss. Pantheists, for example, believe the physical universe itself has a spiritual dimension. The last chapter adds a final word about the task of philosophy.

"Further Reading" lists fifty books and articles I recommend for someone who wants to begin studying philosophy and includes books by everyone discussed in the text. "Teasers" deals with problems of personal identity and the identity of other people. You can get through life without confronting these problems, but you may find them interesting and fun.

Since I am an atheist, you may wonder why this book spends time on whether God exists and what this would signify if it were true. The answer is that I was raised in a country, the United States, that may be the most Christian in the world, know what real faith is like from the inside, respect those who still have it, feel they deserve a serious discussion of their beliefs, and recognize that, for a believer, God becomes the organizing concept of one's whole life.

I am not much concerned with those who have only conventional "faith". Søren Kierkegaard said real faith meant being "willing to give up whatever it is that you love more than God". He added that claiming you have faith without serious prayer is like claiming you can live without breathing, and that real prayer is not merely repeating words at church but seeking communion with God. It is up to you to assess whether your faith qualifies.

A final word to those who may object that I have not been even-handed. Philosophy texts are even-handed but rarely exciting. My goal is to excite you, the reader, by exhibiting what philosophy can do to and for you. James Joyce believed people have to forge in the "smithies of their souls" the person each of us can become. This book is based on what philosophy did to me. I would probably not be very interesting if I did not have passionate convictions about the most important problems that exist. You may come to believe that very different convictions can pass the test of reason.

What is Good?

DOES A MORAL REALITY TELL US WHAT IS GOOD?

After the age of twelve I lost Catholicism as the foundation of my commitment to humane ideals. Indeed, one of my objections to Catholicism was that its rules, such as the ban on contraception, were not always humane. For the next forty years I searched for something that would elevate humane ideals into principles everyone ought to respect, whether inclined to or not. Western philosophy offered five candidates: moral realism, moral language, nature, a universal maxim, and "the market".

Those who posit a moral reality believe there are moral "properties" that illuminate what we ought to do. These belong to an entity—such as God, or the world of forms proposed by the great Greek philosopher Plato—that exists beyond the physical universe, or to states of affairs within the physical universe.

Those who believe in the latter are divided into two camps: those who think moral properties belong to a kind of behavior (such as enjoying the pleasures of friendship), and those who think they belong to people as a trait, which we must use to explain their behavior. (For example, Hitler's wickedness influenced what he did.)

Living in a shared physical universe confers objectivity on science. What if what exists for you did not exist for me—for example, if you saw Mars with two moons and I with three? The advantage moral realism would confer is that my ideals, or those of someone else, would have objective status. If others had contrary ideals it would simply mean they were out of touch with moral reality, just as a flat-Earther is out of touch with the real physical universe.

The oldest appeal to a moral reality is to claim there is an all-powerful god who is morally perfect and issues commandments that tell us what is right and what is wrong. But would we accept his opinion if he were an all-powerful devil? Power is not the same as goodness: if Hitler had created the universe and resurrected himself from the dead, I would still reject his ethics. To accept a god's ethics, I must judge that this god is not only all-powerful but also benevolent. Indeed, why not just benevolent? What has power got to do with it? But that shows I have already made up my mind about what is good: goodness is benevolence.

An all-powerful god also poses the problem of the existence of evil. If this god is omnipotent, why does he tolerate so much evil in the world? Even if you solve this problem it does nothing to alter the facts: you had made up your mind about good and evil prior to judging him. Plato put this with elegant simplicity: do we accept a god's laws because they are good, or simply because they are a god's?

Plato, who lived from 428 to 348 BC, argued that beyond the physical universe there existed a world of "forms", each of which possessed the property of perfection—that is, each was perfect of its kind. The "form" of human society was the morally perfect human society, and we could look to this to know the truth about justice. Its perfection was absolute and thus it was too perfect to exist in the physical world of *actual* human societies.

Although no human society can be perfectly just, we must, of course, strive to approach justice as closely as possible. Think of the concept of a straight line. It is the only line that catches what a straight line really is: the shortest distance between two points. Every line I draw on paper will fall somewhat short of its perfect straightness.

Things that exist beyond the physical world—and so cannot be perceived by the senses—we refer to as transcendental, or non-natural. Plato's brand of moral realism proposes a non-natural entity—the "form" of human society—that possesses a non-natural property, perfect justice.

Plato argued that for every class of object we must posit a general idea in our minds. We can all distinguish chairs from tables. Therefore, even those two chairs that resemble one another least—say, a Victorian rocker and an aluminum lawn chair—must have more in common with one another than either has with a table. The general idea of chair must be truly general, and that means it cannot be reduced to a sense image. It must, for example, be broad enough to include both red and black chairs. To be a sense image it would have to have a particular color, and that would mean it could not do its job of allowing all chairs into its class.

Plato's reasons for thinking each general idea must have an independently existing counterpart (that is, that there must be a "form" of chair as well as a general idea of chair in our minds) are not made explicit. Perhaps since we, as physical creatures, are imperfect and the general idea is perfect, we must get it from contemplating something outside ourselves. However, the external existence of the forms is not too important for ethics: they would hold the key to the perfect state of things even if they were mental entities. Either we can read off their contents and learn what justice really is or we cannot. Plato's method of opposing the perfect and the anti-perfect is called dialectic.

We find that all societies that exist in the physical world share a common defect: they are ordered by politics, which always means a struggle for power and enshrines the principle "might makes right". This is the opposite of justice and suggests what we will find in the form of human society: the struggle for power will have been eliminated and a new, uncontaminated ordering principle will have taken its place. In other words, whatever ordering principle remains after we have eliminated the struggle for power will be true justice and tell us what justice really is.

When we perform the conceptual experiment of eliminating all the sources of the struggle for power, we are forced to imagine a kind of society in which the rulers live simply and have families in common, so no one can suspect them of greed or nepotism. Rulers are selected on merit, for their wisdom and virtue, by an apolitical institution—

namely, the education system. The principle of merit dictates that everyone in the population, whatever the circumstances of their birth, has an equal chance in the education system, and the system must discover and enhance their best talent. When they leave there must be meaningful work for them to do—work that gives them a sense of inner worth so they will not seek "that good opinion they lack of themselves" from others. Empty people have a terrible need to feel larger than life: they seek to be applauded and worshipped and this is the root of the lust for power. Finally, there must be no extremes of rich and poor.

The citizens must believe in the "myth of the metals"— that is, have the proper mores. They must not have commercial values (worship the successful entrepreneur), militarist values (worship the man on horseback), or populist values (admire whoever can attract the applause of the mob). Rather, they must believe that wisdom and virtue are the mark of a ruler and that all social roles should go to those who have the appropriate competence and virtues.

We now realize we have the ordering principle we were looking for, the criterion of justice that orders society without defect. It is a full appreciation of the notion of merit: no racism, no sexism, no aristocracy of birth, no rich and poor, no human life wasted, no lack of a sense of self-worth.

Has Plato found a non-partisan test of goodness, one that confers objective status on his ideal society? That depends on whether the criterion "eliminate all competitions that confer power" is both impartial and sufficiently informative. Few would want an unmitigated struggle for power that would plunge us into an anarchic state of nature. However, many would want to retain competitions for power or wealth as long as these are governed by rules.

Democrats will want to use elections to select the government unless Plato can give better arguments against this than are found in *The Republic*. If they want to play his game, they will argue that elections are not ideal but the closest possible approach to perfection possible in the physical world. Free-marketers will argue that capitalism is the best possible approach to a just allocation of resources between the industrious and the lazy.

Those who are committed to the totality of humane ideals will note that humanism includes the creation of beauty, delight in diversity, and tolerance among those who differ. The criterion of eliminating the struggle for power is too narrow to address these great goods. Plato treats them as devoid of intrinsic value: he merely assesses them as means, either productive or counterproductive, to other goods. He censors the arts and forbids foreign travel because he believes artistic freedom and alien influences will corrupt the masses.

Much can be said for and against Plato on these issues but the point is this: the criterion of goodness he finds in the "form" of human society has been shown to be partisan. This means no one will profit from adding to their arguments a tag such as: "And besides this, my views are in accord with the form." The criterion can be legitimately debated and need not be accorded the role of an impartial arbiter.

The positing of a moral reality is seen to be futile. Plato has not solved a dilemma that afflicts all attempts to establish a higher court of appeal that will tell us whose ideals have objective status. To be non-partisan the court must be chosen according to a criterion broad enough to include both the humane and anti-humane. But in that case how could its decisions favor the humane over the anti-humane? Either it is non-partisan and helpless, or it can hand down decisions only because of its bias.

G.E. Moore (1873–1958) dragged moral properties down out of the heavens. Like Plato, he believed the property of goodness, or moral perfection, was non-natural and should not be confused with the natural properties we see around us, such as the colour yellow. However, goodness did not belong to "forms" that transcended normal human experience, but rather to certain states of affairs we experience every day (if we are lucky)—such as friendship and beautiful objects.

Moore gives us directions as to how to "cognize" which states of affairs have the property of goodness: we are to contemplate them in isolation from everything else. This will winnow out things that have intrinsic value from things valued merely as means. If you do this with money, you immediately perceive that it has value only in relation to the things it can buy. If you do it with the pleasures of friendship, or the contemplation of beautiful objects, you find that they (and nothing but they) are intrinsically good.

It is immediately apparent that Moore's method of "cognizing" what things are really good is hopelessly subjective: the philosopher Friedrich Nietzsche just needs to say that contemplation of beauty is valuable only if the people in question can actually appreciate beauty (supermen), and friendship is valuable only between creatures worthy of regard (supermen) and not when herd men (the rest of us) are tasteless enough to enjoy the company of one another. So once again the positing of real moral entities is useless.

Worse, Moore's method introduces a confusion because "morally good" is not the name of a property but a label we paste on human behavior. The whole notion of natural states of affairs—those belonging to the physical world—having non-natural properties is suspect. No one would think it sensible to attribute natural properties to a non-natural entity, to say something like "God is yellow".

Moore gives a criterion for distinguishing natural properties from non-natural properties. He says that stripping a natural thing of one of its natural properties does violence to it. Strip a lampshade of yellow, for example, and you have altered its whole material structure. Now, rather than reflecting the wavelength of yellow light, it reflects, say, the wavelength of red light. He contrasts this with the pleasure of friendship, which you can strip of its goodness and leave untouched.

Well, so you can, but isn't that a reason for saying goodness is not an attribute of the object at all but a label human beings paste on some human behavior—one that reflects their assessment of the behaviour's moral rectitude? If people assess differently, what changes is not the object assessed but the people: they are revealed to have traded in their old moral principles for new ones. How we judge what is good is one of *our* properties, not a property of things outside us. We will address what "morally right" means in the next chapter.

Today moral realism is making what I hope will be its last stand. The Cornell School of Philosophy asserts that human behavior has moral properties—or at least human beings do—but, contrary to Moore, its proponents say these are perfectly natural properties—that is, they are moral facts about human actors to which we must appeal in order to explain their behavior.

Nicholas Sturgeon (born 1942) asserts that morality was a cause of the abolition of slavery. He is quite correct. The British Crown used the Royal Navy at great expense to suppress the slave trade, and no national interest of Britain dictated this sacrifice. He asserts that Hitler's morality—or moral depravity—was a cause of his behavior. Who would deny that? If Hitler had not firmly believed in a moral crusade to exterminate the Jews, it is doubtful the Holocaust would have been as horrific as it was. Who would deny that if people are really committed to their moral principles and therefore act on them, these principles influence their behavior?

However, these "moral facts" Sturgeon cites seem to me devoid of moral significance. They are simply psychological traits like any other psychological trait. If, for example, someone is really committed to getting into medical school, he or she will study hard.

The fact that moral principles are one of the traits influencing human behavior needs emphasizing. There have been times when psychologists wanted a "value-free" social science and liked to ignore moral principles in favor of more scientifically respectable traits, such as anxiety, self-image, love and emotional control. Or they smuggled in moral traits but tried to pretend they were something else, such as group feeling, or the nurturing instinct, or even disguised self-interest.

Sturgeon is not happy with internalized moral principles being an important psychological trait. He wants them to be called "moral facts" to distinguish them as something over and above what ought to be called "psychological traits". I am never inclined to forbid people using words in ways that give them pleasure. We will call internalized principles a moral x (fact, value, whatever you will) that influences behavior, and never call it a psychological trait.

As for these two categories, there will be close cases. For example, where do we put honesty? But let us set that aside. What does this naming exercise have to do with ethics, unless we add a distinction between moral facts and moral fictions, unless we reserve the right to say, "Hitler was morally depraved" and deny a Hitler youth the right to say, "Hitler was a moral hero."

Ethics is concerned with whether we can separate moral principles into those with objective status—that is, worthy of regard by all—and those with subjective status—not worthy of regard no matter how passionately you believe in them. The superior causal potency of one moral principle over another does nothing to show that it is objective rather than subjective. If it could be shown that only humane moral principles had the capacity to influence behavior, I would be amazed and pleased. But even that "moral fact" would do nothing to vindicate humane ideals. Imagine the shoe were on the other foot. In the wake of a nuclear war, the only ones left alive are ten Hitler youth. Then only Hitler's principles would have the capacity to influence human behavior.

Sturgeon himself says that his case for moral realism depends on whether a certain set of claims about what is really moral can attract rational agreement. This gives the game away. If we can show this, we do not need moral realism to solve the fundamental problem of ethics. Even then, clear thinking would still forbid us attributing some kind of heightened causal role to the vindicated moral principles.

If we really think that moral facts supplement psychological traits in influencing behavior, we should not confine ourselves to the human species. Chimpanzees have an awareness of personal identity. Unlike cats and dogs, they can recognize themselves in a mirror. They also have internalized moral principles. Two chimpanzees sit at the opposite ends of a collapsible sliding tray, each with food in front of him. If one chimpanzee victimizes the other by pulling the

<blockquote>

IS THERE A COMIC REALITY?

A person with an ordinary sense of humor offers a hilarious witticism. A person of temperate habits becomes a hopeless alcoholic. Someone as dull as they come falls madly in love. Comic facts, alcoholic facts, romantic facts: these causal forces are operating all around us. What social scientist trying to improve social science would have walked down this road? The lure of moral realism is simply that it objectifies morality.

</blockquote>

tray forward to get all the food, the other collapses the tray. (He cannot pull it back.) When a human being intervenes and pushes the tray towards one and away from the other, the victim gets agitated but does not collapse the tray: he knows his fellow chimpanzee is innocent. Have animal psychologists missed a trick? To explain this behavior, must they say that the selfish chimpanzee was really wicked and the fair-minded chimpanzee really good?

Sometimes I believe that what these moral realists are saying is that Hitler's behavior was so wicked that, even taking Hitler's moral principles into account, psychology cannot explain it. I cannot see the problem: there were plenty of other Nazis just as dedicated as Hitler, and who would have done what he did if they had had the power. But even if there is a problem, to say the act was so wicked that "wickedness" must be a causal factor is not enlightening (see box above).

And it is not the only possible choice. Gordon Graham, an Anglican priest, sees a problem with serial killers. What they do is so wicked and they appear so ordinary that psychology cannot explain their crimes. Something external to the self must be at work. It is, of course, the devil. The next step is clear: if the devil exists so must God.

It is time to terminate moral realism while it can still have a decent burial. Every kind of moral realism is bankrupt unless it can tell us how to access the morally real and discover what is good. Compare the masterful attempt of Plato to solve this problem through dialectic with Moore's earnest but hardly convincing cognizing, to Sturgeon's positing moral facts to supplement psychology.

DOES MORAL LANGUAGE TELL US WHAT IS GOOD?

The next candidate to give humane ideals objective status is appeals to language. I was never tempted by this possibility, but during my youth the prestige of thinkers who took it seriously was very high. In addition, we owe it to ourselves to be clear about what moral language means.

I will begin with A. J. Ayer (1910–1989), whose book *Language, Truth, and Logic*, published in 1936, was influential in the English-speaking world for the next twenty years. Ayer held that words such as "right" and "wrong" refer to our emotional experiences; those who followed in his footsteps called themselves emotivists.

The thesis is not that words in our moral vocabulary describe our emotions—this is supposed to be a fallacy—but that they *express* our emotions, like shouting "bravo" at a concert. For example, when we say, "Stealing is wrong", this is equivalent to saying "Stealing ugh", with the "ugh" expressing a sense of repugnance or disgust.

This made ethical propositions too trivial even for Ayer, so he added that they were weak commands as well. They were intended to influence the conduct of others. Shouting bravo at a concert can be catching. Our expression of disgust at the idea of stealing is intended to convey that we want others to find stealing repugnant too, and therefore refrain from doing it.

Ayer says the language of a weak command is "You should not kill"; the language of a strong command is "Thou shalt not kill". Only the former is moral language (it uses the word "should"). Ayer does not explain why moral language cannot accommodate strong commands. (It seems unlikely everyone would be indecisive about their loathing of murder.) I like to think it was because of unease about claiming that moral propositions are commands at all.

It is easy to show they are not. Commands do not actually say that things are good. I might try to shove you out of a room because of a fit of anger with no moral judgment behind it. The command "Leave the room!" is just a verbal shove. If I have psychological ascendancy over you, a command takes much less effort than using physical force.

Many an officer has had to give a command he or she thought stupid or wrong. If commands and moral propositions were the same, it would be logically contradictory to say "Leave the room" and also say "You should not leave the room". But this is not logically contradictory: it is psychologically odd. Unless we give into temptation, we will not try to get other people to do things that are against our moral principles. To do so without some reason is to act at cross-purposes, like saying to someone "Go to medical school" and then saying "but you should not put in an application". The primary reason we do not always do what we think we should, that we issue commands to our body that are contrary to our principles, is that we are too frightened or weak to live up to our ideals.

It is also true that it is rare to hold moral principles unless the acts they commend arouse some sense of approval and those they condemn arouse some sense of disgust. But these are merely signs that to hold a moral principle is to internalize it. It is conceivable that this could be done without emotion, as Immanuel Kant suggested. He thought a purely rational creature would espouse moral principles simply because these principles were dictates of practical reason. Certainly, on occasion all of us act on principles we deeply believe in but may find personally repugnant—for example, a shy person may believe she should always speak out publicly against evil.

What, then, is moral language really about?

I have principles, and on the basis of those principles I make a judgment, and that judgment determines what I label "morally right" and what I label "morally wrong". I believe in racial equality and therefore, when I was a young lecturer in the American South, I judged segregation wicked and labeled participating in the black protest movement morally right. "Morally right" was not the name of a property my participation possessed; it was a label I had pasted on my participation. And when I pasted on this label I did not merely express an emotion. The meaning of the label was that participating in the black protest movement was a duty I was obligated to perform.

But what is it about the label "morally right" that entails a duty? It is because moral goodness has a peculiar significance. All other value labels, such as "beautiful", "pleasant" or "useful", imply no comparisons, unless I make the comparison explicit. That I enjoy steak more than hamburgers must be spelled out. That I put a pleasant evening ahead of purchasing a beautiful object must be spelled out. However, when I label an act morally good I am implying that the act takes priority over all other forms of goodness. There is, therefore, no need for me to say so. That it is wrong to make fun of a friend automatically takes priority over the fact I may enjoy doing so, or that someone would perceive the joke as having artistic merit. This is not to say that the creation of beauty cannot get a moral label: the absence of beauty might drain a good society of much of its goodness for the human beings who live in it. Therefore, we have a duty to provide public art galleries.

Spelling out the meaning of the label leaves open what you paste it on—that is a matter for your moral principles. Moore pasted it on the pleasures of friendship. The meaning is nothing less than that some acts are obligatory. That is a solemn thing: it may cost you your life. Before you say that you hold a moral principle you had better assess whether its roots are so deep as to define who you are.

Some believe we do not need a case for "objectivity" in ethics because simply using moral language is equivalent to attributing objective status to an act. Ludwig Wittgenstein

(1889–1951) called making a case for objectivity a chimera. When asked, "What about someone with anti-humane moral principles?" he said he would simply say they ought to hold humane principles. As for his grounds, he observed that calling something "good" or "right" in an ethical sense was identical to saying it was good in an absolute or objective sense.

I believe we must look at language from the point of view of both the speaker and the person spoken to. Or, if you prefer, when I speak I must at least anticipate what my message will mean to a sane person. I am tempted to tell Nietzsche: "You ought to adopt humane ideals." However, I should ask myself whether I do this to vent my antagonism towards him (in which case perhaps I had better chop wood) or whether I am giving moral advice. Unfortunately, moral advice makes sense only if the person can take it. (Ought implies can.) To tell me I should run a hundred meters in ten seconds makes no sense because I cannot do it.

Under what circumstances could Nietzsche, deeply committed as he is to the ideal of the superman and with a deep loathing of humane moral principles, follow my advice? Only if he lost his mind or memory and, reduced to a rudderless ship, would do whatever I said just because I said it. Does it really make sense to use moral language when it presumes insanity on the part of the hearer? Is Nietzsche supposed to accept a set of duties that may cost him his life on my say-so?

Wittgenstein's contention that we can trust how we talk about ethics seems to me even odder than moral realism. Rush Rhees (1905–1989) endorsed this contention, saying that to speak of our ideals lacking objectivity is meaningless. After all, it is mistaken to say that something exists only if we perceive it to exist. What then is the sense of saying that something is good only if we judge it to be good?

My reply is dual. First, I am not interested in how people talk but in how they would talk if they were reasonable. Secondly, if they were reasonable they would tailor what they said in the light of what exists and what can be known. Language is not the final court of appeal. More fundamental are metaphysics (what exists) and epistemology (what is true). It makes sense to say "God exists" only if you believe he exists, and you should believe he exists only if you have some way of knowing that he does, even if this is only an appeal to faith. Failing that, you should purge your vocabulary of God talk (see box opposite).

Rush Rhees is correct on one point: it seems foolish to say that things exist only if we perceive they do. But why is that? It's because we live in a shared physical universe. Imagine we didn't, that every person lived alone in his or her own physical universe but could communicate with others by mental telepathy. I get a message from someone saying, "My observations show that heavenly bodies attract one another inversely as the distance cubed." I send back, "That is odd. My observations show it is inversely as the distance

NO SIN OF VERIFICATIONISM
The fact I cannot show that humane ideals have objective status curtails my tongue, but this is not what academic philosophy calls verificationism: it has nothing to do with the meaning of ethical propositions. Despite my failure, the meaning stays the same: all mankind has a duty to be humane. It is just that saying that—to Nietzsche, for example—is foolish if I have no case with which to back it up. Once again, we know what "God exists" means, independently of whether we can verify God's existence. Nonetheless, setting "faith" aside, it is foolish to tell people that God exists unless you can offer some kind of argument that he does.

squared." He replies, "Well, we do live in separate universes, and I guess the kind of gravity that exists for you is different from the kind of gravity that exists for me."

Under these circumstances, saying the kind of gravity that exists is contingent on our personal experiences is perfectly sensible. You may quibble that both of us agree on something: the kinds of gravity existing in our two physical universes. Well, Nietzsche and I can agree on what conduct is sensible given our differing moral principles. What we cannot agree on is what all people should do.

The question then arises: Do we live in a shared moral universe? The answer is no, because there is no such thing as a moral reality. We are each trapped in our own private moral universe, one delineated by the particular moral principles each of us holds. If two moral universes coincide, that is all well and good. But if they do not, as with those of Nietzsche and a humane person, the lack of a shared moral universe must be made good by epistemology. You must have an argument ready to show that some higher court of appeal renders an impartial judgment that your moral universe and its principles are worthy of regard from those who do not share them—those who have anti-humane moral principles. As our critique of Plato indicated, we have reason to believe no such case is valid.

The absence of a shared moral universe, and of a case that bridges the gap between unshared ones, makes it sensible to defy ordinary language and say that humane acts are obligatory only on those who hold humane principles.

I have found that this conclusion often provokes an endless wrangle over what we have a right to say, and therefore I will close this chapter with something I do not believe to be contentious. The following is meant to show that it is pointless to debate whether the assertion "We ought to behave humanely" entails the equivalent of objectivity.

"We ought to behave humanely" can be divided into two propositions:

(a_1) Humane behavior is a duty people should put above all else;

(b_1) This is so, whether you internalize humane ideals or not.

To this it must be added that a_1 entails the truth of b_1.

However, if someone does not accept a_1, the fact it entails b_1 gives him no reason to act humanely. In addition, I have no case to that effect (beyond the irrelevant entailment). Therefore, if someone who does not hold humane ideals acts humanely simply on my say-so, he is insane.

To this it must be added:

(a_2) I will behave humanely;

(b_2) I will tell Nietzsche that he, too, should behave humanely, but only when he cannot hear me because I would feel so foolish if he could.

DOES ECONOMICS TELL US WHAT IS GOOD?

Despite the obvious flaws of these attempts to objectify humane ideals, I found it almost impossible to gain peace of mind unless I could discover something rational that would put a stamp of approval on them. It took years to face the fact that I was yearning for something that did not exist—and worse, something that could actually do violence to the richness of humane ideals.

Many attempts to justify humane ideals have not only failed but also gradually drained these ideals of much of their content. The villain is the search for objective status. Everyone—including me, until recently—wanted a case that would force all humanity, including Nietzsche, to choose between adopting humane ideals or confessing they were rationally remiss. The fear was that if humane ideals were not "objective"—that is, in accord with reason—they were merely "subjective"—that is, they appealed only to those who admired them—and were therefore as arbitrary as our preferences. The story of the decline of humane ideals begins with the Greeks.

Plato's just society had a rich content. Aristotle (384–322 BC) did an even better job of articulating what a good society is about.

Aristotle says that civil society is more than a market because you can do business with foreigners, more than a military alliance because you can negotiate mutual defense treaties with foreigners, more than marriage ties because you can marry a foreigner, more than physical proximity because two groups can occupy the same city and be divided by hate, and more than abstaining from injury to others because one can be kind to foreigners. The foundation of a civil society is a shared way of life, rich in philosophy, art, sport, amusements and diversity, whose consummation is a sense of personal loss if anyone else suffers the deprivation of non-participation. The society allocates benefits and duties fairly. It develops the full potential of its citizens, unlike dwarfed societies that cultivate only the entrepreneurial (Carthage) or military virtues (Sparta).

In sum, the humane ideal includes six great goods. These are like six balls we must juggle in that there are always trade-offs between them. They are also the foundation of social criticism from a humane point of view. These great goods are: *happiness*, at least to the point where misery is abolished; *justice*, particularly that none are denied the good life because of bad luck; *the creation of beauty*, it being a warning sign if this is limited to a professional élite; *the pursuit of truth*, not indoctrination by government or deception by the market; *tolerance of diversity* or, better, *delight in diversity*, rather than a stale conformity; and *the perfection of human nature*, rather than people being socialized into mere consumers, soldiers or spectators.

Aristotle justified his ideal by proposing that humankind had a "final cause". This amounted to a hypothesis: human nature is such that only in the good society will people maximize their sense of psychic wellbeing, or "eudaimonia".

You can rebut this by saying whether or not a way of life maximizes psychic wellbeing is irrelevant. Moral goodness is not the same as medical goodness or psychological health. If there were evidence that Hitler youth enjoyed the apex of psychic wellbeing, I would not imitate their behavior: better a bit of neurosis and humane behavior. Duty sometimes demands the sacrifice of mental health. Young Vietnamese cut throats for twenty-five years to free their nation from foreign rule. I doubt they emerged totally sane.

But the point is that Aristotle tried to justify humane ideals as a package, an image of a full-blooded good society. When his attempt failed, his method did not set aside even one of the six great goods.

Modern thinkers rejected the ancients' cases for objectivity, but they wanted to justify what humane values they could. Sadly, their attempts did not leave humane ideals intact but progressively diminished them.

John Locke (1632–1704) turned to nature, or the intentions of God revealed in nature. Nature, he said, gives each person the instinct for self-preservation, which shows that suicide is wrong. Murder is also wrong, because you have no right to do to another what he has no right to do to himself. Since murder is wrong, slavery is wrong—under the law the slave's life was forfeit to his master—unless the slave is a culprit worthy of the death penalty. Nature gives each of us parental instincts, which shows that nurturing children is a duty. Nature gives us free will, which shows that government should be based on consent. Our only natural possession is our labor, which means that property is legitimate only when our labor is mixed with it.

Note what has happened. No longer do we have a humane society complete with the six great goods, but rather a set of moral principles that forbids brutality, tyranny, and unjust acquisition of property. The mode of justification has cost humane ideals their emphasis on the pursuit of truth, beauty, and delight in diversity, the rejection of materialism as an assault on human potential, and the presumption that all have a right to access the richness of a shared way of life. The humane ideal has shrunk and objective status has still not been realized.

As John Stuart Mill (1806–1873) pointed out, nature no more tells us what is good than God does. Nature shows examples of great cruelty that no one would want to imitate. We make up our minds about what is good quite independently of nature, and then we select from nature things we like, such as people having free will, and ignore what we dislike, such as nature's total indifference to human suffering.

It is not Locke's fault, but the use of nature as a justification still today perverts moral argument (gays are unnatural) and even misleads us about why we want to preserve nature (save the environment). No one wants to save all things in the environment. We always pick and choose: for example, we exterminate possums to save trees and kill weeds to beautify our lawns. Nature possesses no intrinsic property of goodness; it is we who label it good and find it beautiful, rich in variety, and so forth.

Mill did not give up the quest for objectivity, but he believed he could prove the objective status of only one supreme maxim: promote the greatest happiness of the greatest number. So now we are down to one ball, happiness.

Mill and his followers, the utilitarians, twisted and turned to show that we could use this to conjure up all the others. They suggested we all prefer beauty when we have a choice (do we?); that the general happiness usually dictates justice (not always: we may enhance the happiness of most by sacrificing one innocent person); and that free debate usually promotes truth (but truth does not always make us happy).

Mill's proof was fallacious: it leapt from saying that each of us values our own happiness to the conclusion that I should value the happiness of others as much as my own. This is like saying that if I value money I am indifferent as to whether it is my money or your money.

William James (1842–1910) thought he could find a proof in the failure of all previous proofs. Since no one could make a case for using forms or nature or happiness to rank human demands, there was, James said, one unconditional commandment: Satisfy as many human demands as possible without regard for whose they are or what they are for.

The argument is bankrupt: if there is no case for objectivity, we are all free to evaluate demands as we wish. I can put my demands ahead of yours, European demands ahead of African, male demands ahead of female. Note that even James puts human demands ahead of those of animals.

And what has his attempt to make a case for objectivity left us with? Sadism is fine if you have enough money to hire a willing victim. If advertisers can make us demand endless possessions, maximizing possessions is good.

James left ethics ready for a takeover from economics, for who knows how to fulfill as many human demands as possible at the least cost if not the economists? Some of them have ceded moral authority to the market, claiming the law of supply and demand is the best judge of what people like and interference is impiety.

Even economists who believe the market can malfunction have internalized James' maxim. For example, Alan Blinder (born 1945) is a thoroughly good man who does not like what the market does to the poor or the environment. He begins by stating a principle he considers beyond dispute: "More is better than less. If the people did not want what the market produces, unloved items would find no market." Well, what objection could James have to that?

In summary, the lust for something that can tell us humane ideals are objective has virtually destroyed these ideals. We slid from the rich image of the good society to a scatter of moral principles, and finally to a supreme maxim bankrupt of anything except naked demands. Our final refuge has become the market. Why is that so disastrous?

Ethics must not bow to the market, whether the market is unfettered or repaired by some tinkering. There are six great tensions between the market and the good society.

One, market calculations create conflicts between self-interest and justice. For example, a poor widow wants to rent her rooms to Americans irrespective of race, but she knows that one black male in three is a convicted felon and she cannot afford to take the chance of an unsuitable tenant.

Two, the market cannot accommodate the full spectrum of human activity that promotes excellence and pleasure. If what I wish to do to enhance my life is to gain a living in the market, others must be willing to pay for whatever results in terms of goods or services or performances. Most people, particularly industrial or service workers, earn a living by doing many things they would never freely choose to do.

Three, all should have access to the good life but the market makes access to basic goods into prizes won in a competition by those who succeed.

Four, these competitions often have results no one desires. Competition for homes in good neighborhoods with good schools may give private advantage but inflate the cost of homes, mobilize sentiment against local property taxes, and leave less funding for schools.

Five, the moral glue of the social order is civic virtue. Civic virtue is based on fellow feeling, or the conviction that everyone should have access to the good life. The market does not promote the sense of personal security that allows fellow feeling to flourish. Instead, citizens feel totally dependent on their own resources to protect themselves and their families against misfortune, and so are reluctant to pay taxes that divert some of their resources to provide for others.

Six, the market has created its own powerful institutions that willingly acknowledge no moral limits—for example, corporations that encourage poor nations to buy arms or their people to smoke tobacco.

Some advocates of humane ideals did not forget what they were all about. However, broadly based social criticism tended to gravitate outside philosophy. Novelists such as F. Scott Fitzgerald (1896–1940), John Barth (born 1930), and Aldous Huxley (1894–1963) weighed in. Huxley in particular, in his utopian novel *Island*, used all six great goods to criticize industrial society. He criticized its materialism as demeaning and, worse, as discouraging the mystical experience. He criticized militarism, injustices, misery, intolerance, and systematic lack of regard for truth.

Social science offered social critics, the best being Thorstein Veblen (1857–1929), Ruth Benedict (1887–1948) and R.H. Tawney (1880–1962). These critics attacked the excesses of materialism, which can debase almost anything: human dignity (for example, trophy wives), art (painting to please the critics), truth (writing profitable trivia), tolerance (the undeserving poor), and happiness (the literature shows that recent economic miracles actually reduced the level of happiness).

In recent years some thinkers have restored humanity to philosophy. Karl Popper (1902–1993) never wavered in his social analysis. Marxist and feminist influences energized Kai Nielsen (born 1926), Thomas Pogge (born 1953) and others' search for justice on the international scene.

The attempt by John Rawls (1921–2002) to resurrect humane ideals deserves comment. Rawls constructed a general theory of justice, but he focused mainly on justice rather than on the rich image of a humane society. I suspect this was because he thought he could give non-partisan arguments for justice, but had none to give for other great goods.

Rawls imagines people in an original position, motivated purely by self-interest, who must choose what kind of society they want to live in. There is a veil of ignorance: they are denied knowledge of any special talent or status they may attain—male or female, rich or poor, bright or stupid, best off or worst off. They are not allowed to choose any kind of society; rather they must remain ignorant of whether it will be Catholic or Islamic, medieval or modern, affluent or poor. They are limited to choosing how it allocates benefits to those on various levels of the social hierarchy (and presumably whether or not there is to be any social hierarchy). It is a mystery why the choices of selfish and ignorant people would lend moral dignity to the result of their choices. The question is not posed, perhaps because Rawls liked the outcome.

It is assumed that these people will choose a society in which the worst off are not badly off—here comes the welfare state—and those at the top will get more only if their gains also result in an improvement in the lot of those at the bottom. Certainly, people on the liberal left would endorse all this.

Rawls does good work spelling out a liberal concept of social justice, although his mechanism corrupts what I believe to be the strongest argument for a safety net under the unfortunate—namely, that bad luck should not doom people to a bad life—into enlightened self-interest. Nonetheless, the detail is impressive.

However, as far as lending these principles of justice some sort of impartiality, Rawls' mechanism is a failure. It stacks the deck against élitists by obliterating the differences between people that they use to justify discriminating in favor of some over others: namely, that they are supermen rather than herd men, white rather than black. In addition, the choice that people in the original position are supposed to make is distorted by wishful thinking. And why *people* in the original position? One person would be enough in that they all think the same.

Assume you must participate in a lottery with money prizes and your prize is dictated purely by chance. In addition, you are ignorant of the amount of the prizes because both the total amount of money available and how it is allocated are unknown. There may be enough money, even if the prizes are very unequal, to give everyone affluence. Or there may be so little that, even if the prizes are equal, or indeed even if one person gets the full amount, everyone will be sunk in misery.

With only this information, you must choose how the total sum should be divided up. I have made the prizes money to simplify things. They could just as easily have been all the social benefits that contribute to good living. Not only your fate but that of your spouse and children is utterly dependent on your choice.

Faced with this kind of uncertainty, the most common reaction of a person in the original position might be to jump off a bridge, and this would generate no theory of justice at all. Better to maximize the chance of salvaging something. The most rational option would be this: I decide there will be as many prizes as possible that are large enough to give me a tolerable life—that is, one clearly above the level where my family and I all wish we were dead.

Note that this dictates no particular distribution of the money. If there is very little, it may all have to go to one person to satisfy the condition: better to have one chance in a million of a tolerable life than none at all. If there is a lot everyone may get enough, even if there are huge discrepancies between the prizes. If there is just enough to give everyone a tolerable life, the prizes will all have to be equal. If there is a surplus beyond that, it could be

VEIL OF IGNORANCE

You can play some pleasant games with the veil of ignorance. If you want a prohibition against abortion, make me ignorant of whether, when society pops into existence, I have already been born or am still a fetus. If you want a prohibition against the use of higher animals for medical experiments, make me ignorant about whether I will be a human being or a captive chimpanzee. In other words, by making me selectively ignorant in a way that binds my fate to that of another person or creature, you can get me to wish both of us well. So what?

distributed as unequally as you like, depending on whether you are a gambler.

What is the value for ethics of this strange device? It says nothing about how social goods should be distributed so it has no implications for justice. At best it shows this: if I am ignorant about how I will fare versus the others, and we are all in the same boat, then personal self-interest may dictate outcomes beneficial to others. Faced with this thundering truism, I believe humanists and racists and Nietzsche would all say, "Well, I guess that is so," and go back to some sort of meaningful debate about their conflicting ideals (see box above).

Rawls is better than William James. Nonetheless, as usual the lust for some sort of impartial derivation of humane ideals gives preferment to a fragment of their content. And the case is defective anyway.

Searching for something other than ourselves to tell us what is good is futile. In fact, setting aside Plato and Aristotle, every attempt to locate that magic talisman has been worse than futile, doing violence to the very ideals we want to defend. The lust for objectivity in ethics is counterproductive. Step by step, it drains humane ideals of their rich content and leaves a pallid residue: whatever is *supposed* to be objective. It paves the way for the final ignominy, a takeover of ethics by economics.

DO WE TELL OURSELVES
WHAT IS GOOD?

If neither God nor a moral reality nor nature nor anything else can tell me what is right and wrong, if I must find the answer myself, does this mean my ideals are merely "subjective"? Do they have no higher status in the light of truth that those of a Nietzsche who despises ordinary people as "herd men"?

Before we worry about defending our ideals, we should make sure what ideals we really hold. Without something or someone else to tell us, how do we know what is good and what is evil? Each of us must do what every thinking human being has done both before and after the rise of philosophy: not just accept whatever most people in our culture happen to believe but read widely, survey the diversity of ideals that human beings have developed, try to put ourselves in their shoes and reflect—then look within and ask certain questions.

Ask yourself, "What kind of behavior overwhelms me with moral revulsion?" Reading *The Diary of Anne Frank* and identifying with the people that Hitler systematically exterminated simply because they were Jewish awakens *my* moral revulsion.

Ask yourself: "Do I have a friend who lives the kind of life I admire?" "What kind of ideals do I want to see my children employ to give substance to their lives?" "Who most excites my moral admiration, someone like Eugene Victor Debs, who dedicated his life to improving the lot of ordinary working people, or my boss, who never looks beyond acquiring a bigger car and a bigger house?"

All these questions are ways of clarifying what you hold dear, and isolating the deepest moral principles to which you are committed. All are different ways of asking the question: "Who am I?" If you look within and find nothing, that tells you something about yourself. It is not philosophy that is responsible, it is your poverty of soul.

Our ideals, or lack of them, define who we are. And for some people living up to their ideal self is rather important. This provides the only answer that can be given to the question "Why be moral?" You are moral because you are deeply committed to principles that demand humane behavior—moral principles that label humane acts as duties that take priority over what you desire. Failure to live up that code exacts a price: you must admit you are not a humane person after all. You have forfeited your membership of the legion of men and women who, sometimes at the cost of their lives, have tried to build good societies throughout the ages.

Reflect for a moment. You are tempted to do something wrong. You find yourself alone with your best friend's spouse, both of you mildly intoxicated and sexually in tune. What would most help you resist temptation: thoughts about a moral reality, the fact that moral grammar is categorical, reverence for nature, or asking yourself, "What sort of person am I?" Of course fearing that God would roast you in hell for all eternity would also be a disincentive, but that rests on an illusion.

Feodor Dostoyevsky (1821–1881) in *The Brothers Karamazov* says that if God is dead anything is allowable. God, for him, was the only source of knowledge of the good. If the good was not objective, all moral ideals were trivialized: they collapsed into the category of mere whim or desire. Risking one's life to pull a child out of the path of an oncoming car became indistinguishable from van Gogh's mad whim to cut off his ear.

Dostoyevsky believed, in other words, that ethical skepticism entails nihilism, in the sense that it becomes irrational to take duties seriously, both duties in general and humane ideals in particular. We may be passionately committed to principles that tell us we should act humanely, but the message of those principles is deceptive: they are like hallucinations.

This logically incoherent argument should be labeled "the nihilist fallacy". Commitment to a moral principle is a commitment to a duty, and even if it is a self-imposed duty it is far more serious than a mere preference for one soft drink over another. In the absence of an ethical truth test of some sort, a humane person cannot tell Nietzsche he ought to accept humane ideals. However, to say that we ourselves ought to abandon humane ideals is to claim more than that they lack objective status: it is to claim that they have *subjective* status and we should discount them as if they were hallucinations.

But why do we discount a hallucination? Because it has failed a truth test: it is deceptive about something. We see an oasis in the desert, and when we get there and try to drink we get a mouth full of sand. If there is no test of objectivity in ethics, humane ideals can neither pass nor fail. There is no test to fail.

What are our humane ideals supposed to be deceptive about? They are not deceptive about our deepest selves. In the absence of objectivity, there is also no such thing as subjectivity. A self-imposed duty to be humane may seem worthless to the anti-humane, but for us it is worth precisely what it is worth to us. That may be a great deal. It may demand that we lay down our lives to avoid anti-humane consequences. To do otherwise would go against our principles.

If humane ideals did have objective status, what would that entail? It would mean we could give reasons for humane ideals that were valid for all humankind—Nietzsche as well as ourselves. We can hardly tell Nietzsche to adopt humane ideals simply because they take the welfare of all human beings into account, any more than he can tell us to adopt his ideals simply because they favor an élite of creative geniuses over herd men. Both of us would be giving purely partisan reasons for our ideals.

You need non-partisan reasons. These would have to somehow bridge the divide between Nietzsche and myself, the gulf between the humane and the anti-humane, so they would have to be neutral, neither distinctively humane nor distinctively élitist. That is why "neutral" concepts such as nature (Locke) and psychological health (Aristotle) are appealing.

But do we really want to substitute neutral reasons for calling actions good, in place of partisan humane reasons? If you justify humane behavior as the dictates of nature, or as what perfects man, you have to mean it. You cannot tell your opponents that humane ideals ought to be accepted because they are in accord with nature without adopting that reason for yourself, otherwise your opponent could accuse you of bad faith: "I thought you told me the real reason for accepting these ideals was that they were in accord with nature. Do you mean to tell me that is not your reason?"

Your opponent would be quite correct. If that is the right way to reason about what is good, it cannot be set aside, like a best room used only for company. You must live in it too. This extracts a heavy price: you can no longer give your true reasons for the ideals you accept—namely, that you hold them precisely because they are humane.

In *Les Misérables* Victor Hugo (1802–1885) introduces us to Sister Simplice, a nun who has always believed that lying is an absolute form of evil and has taken her name from Simplice of Sicily, who was martyred rather than lie about her place of birth. Hugo shows Sister Simplice resisting the temptation to tell lies out of kindness. She meets Jean Valjean, a former criminal, and recognizes his essential goodness. Javert, a detective, approaches her. He wants to arrest Jean Valjean for a trivial offense and restore him to the horrors of the galleys.

Although Sister Simplice has never told a lie, to save Jean Valjean she lies "twice in succession, without hesitation, promptly, as a person does when sacrificing herself". Victor Hugo adds, "O sainted maid! You left this world many years ago; you have rejoined your sisters, the virgins, and your brothers, the angels, in the light; may this lie be counted to your credit in paradise."

Hugo wants to say that the sister's benevolent lie was right precisely because it was benevolent. It is hard to see him settling for something else. Here we see the darker side of ethical objectivity fully revealed. It forces us to set aside the reasons we want to give for the goodness of our acts and substitute the kind of reasons it puts into our mouth—for example, that it is right to tell a benevolent lie because this is in accord with nature.

Imagine that Hugo had added a bit to his account. Sister Simplice tells her lie, Javert leaves, and she says to Jean Valjean, "I see that you are surprised. But over the last few months I have been thinking about the purposes of nature. I now see that the natural purpose of communication does not require that we always convey our thoughts accurately. Just as there are exceptions to everything in nature, we can make exceptions here."

Or she says, "I have been reading Aristotle lately and thinking about what perfects human nature in the sense of maximizing psychological health. I have decided that the tension caused by always telling the truth is too great for human beings to bear and that I am no exception. At any rate, be assured that I did not allow any thought of benevolence to enter in while I thought out this question."

Humanism is not simply a set of conclusions about what is right. It has its reasons. If we are honest, we will admit that neutral concepts thrust upon us a whole range of reasons we do not really care about, to the detriment of reasons we really do care about. The latter may seem "subjective" but they are our own.

Ask yourself this: would you really give up humane ideals if they were not in accord with nature, or if less humane behavior would contribute to your psychological health? (Remember the young man who fought to free his nation of foreigners.) If so, I can only assume you care more about nature than you do about the humane content of your ideals. (For a supposed exception, see box on next page.)

Humane ideals are a rich mixture of six great goods, which include not only the promotion of happiness but also justice, the creation of beauty, the pursuit of truth, tolerance of diversity, and guarding human potential from degradation by materialism and militarism. When we face the fact that these ideals are forged deep within ourselves, rather than handed to us by something external, we preserve them intact.

GOD AS A REASON

A religious person may say they put God ahead of their own distinctive reasons for calling acts right or wrong. They are happy to adopt as their reason: "God wills it." Imagine that God were like the god of the Old Testament and commanded a revenge ethic. You may think it to your credit if you set your humane ideals aside in favor of God's will but I do not. God is still all-powerful but he is no longer benevolent. Might does not make right. I might obey him out of fear but I would have no sense of doing the right thing.

The lust for objective status compromises our reasons for being humane. At best, one great good is salvaged as "objective" and used to rank all the others and render them peripheral. When John Stuart Mill used happiness to rank all the other goods, he had to say that justice, beauty and truth were worthwhile only if they contributed to the general happiness. When Immanuel Kant invoked justice, he said that punishment should never be mitigated, even if mitigation would contribute to the reformation of the criminal.

There are trade-offs between justice and benevolence, between being kind and telling the truth, between devoting resources to the creation of beauty and alleviating misery. Every humane society has evolved its own peculiar weighting of the great goods in various situations. Be happy if you can live with the solutions that have evolved.

CAN WE DEBATE WHAT IS GOOD?

Does abandoning objective status mean we can do nothing to defend humane ideals in the light of reason? It took me almost fifty years to puzzle out this question. You must judge whether my answer is sound or just a soporific.

A defense of humane ideals that is worth having must be non-partisan, significant, and salvage the ideals' integrity. It must not be biased in favor of humane ideals, must levy prices on anti-humane ideals too heavy to be ignored, and must not put new reasons for being humane into our mouths.

There is an agenda that qualifies on all counts: you challenge your opponents to answer the following questions:

One, do you have a criterion as to what creatures are significant enough to qualify for the circle of moral concern—that is, for just treatment and consideration of their welfare? (The circle of moral concern may include anything, from all living creatures to a fragment of humankind.)

Two, do you have a criterion of justice that allocates boons and ills according to merits or desserts? (It need do so only within the circle of moral concern.)

Three, are you willing to universalize the above? Whenever you label something or someone as right or worthy of regard or just, are you willing to give reasons until you reach your basic ideals or first principles, and then stand by those principles with logical consistency?

Four, are you willing to face up to the consequences of your ideals in practise?

Five, are you willing to show that your ideals can serve as the ordering principle of a human society, one capable of organizing people into a social order?

Six, are you willing to show that your ideals do not require you to ignore any truth revealed by science—usually social science, but also biological and natural science?

This agenda is non-partisan. I know of no moral or political advocate of historical importance, whether humane or anti-humane, who did not accept every one of its items. Nietzsche is often called an amoralist. However, he had a criterion that certified supermen alone as worthy of regard; he believed in justice as fairness for supermen; he would be willing to praise anyone, herd man or superman, who promoted the hegemony of supermen; he did not flinch from the suffering his ideals entailed; and he thought his ideals could be used to order a cosmopolitan Europe.

The prices of forfeiting items on the agenda are too great for anyone to pay. I call these the first prices. The prices you may incur when you debate over items are heavy enough to be demoralizing. I call these the second prices. Even they are non-partisan, in that they are imposed by logic or science.

For example, every advocate clarifies who merits moral concern (item one): if they didn't they could not even launch their morality (first price). They must clarify whether their ideals are humane (include all humankind as worthy), or ultra-humane (include animals as well), or racist (whites only), or classist (no deserving poor), or nationalist (my nation over all), or aesthetic élitist (supermen only). Once they have done that, logic or science may show that their criterion is arbitrary in its own terms. If we classify humans as counting because they have a sense of right and wrong, how is it logically possible to exclude chimpanzees, who have shown they too have a sense of right and wrong (second price)?

The price of having no criterion of justice (item two) is that you must tell people, even those you value highly, that you do not care whether or not they suffer arbitrarily (first price). After you have stated your criterion, social science may show that hidden among those you dismiss as herd men are people who, if they had a chance, would qualify as supermen (second price).

Refusing to give logically consistent reasons for your moral assessments (item three) means you must ask people to accept your ideals without comprehending them. Imagine that you praise the military virtues one day and refuse to praise someone who practices them the next. People then have the right to refuse to espouse your ideals on the grounds that they are baffled as to just what kind of conduct those ideals recommend (first price). When racists attempt to be logically consistent about their ideals, they are trapped into declaring that certain personal traits are irrelevant to assessing people. Those same traits are ones that no one, including racists, consistently omit when assessing people (second price).

Those who will not face up to the consequences of their ideals (item four) pay the heaviest possible price: it will be clear to all that they must entertain illusions to believe in their ideals (first price). If we face up to consequences we may have to grant that, in a poor nation, the welfare state starves infrastructure and impedes the economic development that could eliminate poverty (second price).

If you refuse to provide a coherent scenario as to how your ideals could be used to organize a human society (item five), others have a right to suggest that you suspect your ideals are unworkable (first price). In the next chapter we will take up the meritocracy thesis. This argues that whenever humane-egalitarian ideals are used to organize a human society, they produce a society dominated by a genetic élite of the kind egalitarians loathe (second price).

Finally, unless you argue that evading the truths of science is not a prerequisite for accepting your ideals (item six), you give all rational human beings good reason to reject them (first price). On the other hand, if you argue that your ideals are consistent with science, you may reveal that you do not understand what a particular science, such as the theory of evolution, is all about (second price).

Note how non-partisan the agenda is. If you review the second prices, you will see that until logic and science pass a verdict humane ideals are just as much at risk as anti-humane ideals. After all, there are plenty of people, such as Nietzsche, who look within and find a commitment to anti-humane ideals. They can use our agenda against our ideals, just as we can use it against theirs.

Those of you familiar with the naturalistic fallacy will know it asserts there is no logical bridge between fact and values. There are subtle exceptions, but none affect the conclusion that the facts of science cannot logically entail any proposition that actually tells us what we ought to do.

There may seem to be a mystery about our agenda: if there is no logical bridge between facts and values, how can those who hold certain values be so vulnerable to scientific evidence? The answer is that certain bridges are built not when we go from facts to values, but when we go from values to facts. And while the connections are not those of logical entailments, they are so essential to a particular morality that if logic or science invalidates them the morality is crippled.

Value inclinations are not enough to give anyone a morality they can live by. You must link your values to the real world by way of certain factual propositions, and these "connecting propositions" can be tested against evidence. Marx may have had a proclivity to admire the proletariat, but he did not just run through the streets shouting, "I love workers." This would have put him on a par with someone who says, "I hate my boss"—a sentiment significant to that person perhaps, but of little interest to the rest of us.

We may meet a Frenchman who goes berserk every time he hears the word "camel", an interesting quirk but not easy to convert into moral principles. What Marx did was clothe his value proclivity in significance by elaborating a theory of history that gave the working class a crucial role in progress toward a humane society. That puts him at the mercy of evidence.

Although racists may hate black people and that may be the psychological basis of their ideology, they do not just run through the streets shouting, "I hate the color black." To turn their hatred into more than a quirk, they must connect it to the world by assumptions about genetics (race-mixing debases the offspring), history (if blacks only had existed, the human race would never have become literate), human potential (black immigrants will always be a burden on the public purse), and so forth.

Lest we feel too smug, our fellow feeling for humankind is also a mere proclivity until it is connected to the larger world, and it is up to us to make sure we assert no claims that cannot survive the tests of logic and evidence. Just as we demand logical consistency and an honest appraisal of evidence from our opponents, we must make sure our ideals measure up. If you believe in your ideals only because you think every worker is noble, or every woman would be perfect if not debased by the company of men, or all people are basically good, you will not last long in moral debate. Nietzsche asserted that Dickens had never painted a picture of someone who had a "good heart" without describing a fool.

We can now fully appreciate why debating over our agenda is impartial: we are only taking both the humane and anti-humane at their word. They state the bridging propositions that come under attack, and do so because they have no choice.

This method of defending humane-egalitarian ideals, assuming our ideals do prevail, has no darker side. No new overriding reasons for behaving humanely have come to light. We are more secure in our ideals because they have jumped certain hurdles erected by logic and science. We are gratified that our opponents have not been able to jump those hurdles, and at the demoralization this entails. Our opponents must admit they have no coherent criterion for ranking people, no coherent scenario for organizing a human society, and so on. However, we still call humane behavior "good" for the same reason we always have: because it is humane. To show that our ideals can bridge the gap between values and facts is simply to show that these ideals are not mere quirks or proclivities.

The fact we can defend humane ideals in this way gives them no claim to objectivity. The racist cannot defend denying blacks the vote, but nothing has necessarily changed his heart. He may now feel differently about black people, or he may not. He may still loathe people with black skin, shun blacks, and tell his daughter that if she marries one she will be persona non grata. Nietzsche may have to admit that his admired caste society would stifle talent and thwart potential supermen, but he may have no feeling for the sufferings of people who lack the potential to become supermen. He can still advise a creative genius that killing a herd man in an alley is justified if the genius needs money for paint and canvas.

In a word, we have done nothing to show that all people should adopt humane ideals, whether they loathe them or not. We have destroyed certain bridges between anti-humane value proclivities and the real world, but our opponent's anti-humane value proclivities may still be intact. Moral philosophy cannot give you a good heart, and having a good heart is the only reason to cherish humane ideals. The fact we have no case for objectivity has left untouched the integrity of humane ideals. It is ironic that, for that very reason, our victory is qualified.

Let us put our agenda to work against some anti-humane opponents. Classical racists say that all blacks (or Jews) ought to be treated as inferiors. They should be denied rights such as freedom from bondage, the vote, freedom of movement, and freedom to marry any partner who is willing.

The first question we should put to them is "Why?" They then have a choice—namely, to appeal to sheer blackness of skin or name a desirable human trait in which blacks are supposedly deficient.

If racists choose the first option, we can pose a question put by R.M. Hare (1919–2002): What would they say if their own skin turned black, perhaps because we sneaked a pill into their food, or as a result of some pollutant in the water supply? This is, of course, a demand for logical consistency (item three on the agenda).

The first option levies penalties that are subtle but compelling. To say I should be treated badly simply because I am now black may seem to be a heroic willingness to suffer for one's principles. Actually it trivializes one's moral principles. It says I am willing to suffer for an absurdity— namely, that color nullifies personal traits as criteria for assessing human beings. Hitler did not tell his followers they were superior simply because they were white or Aryan. Rather, he told them they were more creative, courageous and commanding than the rest of us.

Imagine a Nazi orator telling his German audience that they deserved to be ruled by Africans just because the two groups had exchanged skin colors. Imagine a book reviewer telling his readers to scorn one book because it has a black cover and buy another because it has a white cover. The next day he tells them the reverse because new editions have reversed the colors. Even racists would stop listening to this book reviewer in favor of one who deigned to discuss plot, character, dialogue and style. If racists grant that it is absurd to ignore the traits of fictional characters when nothing is at stake but a good read, can they contend that we should ignore the traits of real people when the stakes are who has a right to a decent life?

That is why real-world racists choose the second option and assert a correlation between color and despised personal traits. Logic has forced them to enter the real world and assert factual hypotheses. Falsification by evidence follows automatically (item six). We can point to thousands of counter-examples, blacks of genius and talent, ranging from Gordon Parks, the great photographer, composer, author and poet, to Paul Robson, the great Shakespearian actor and multilingual orator, to Thomas Sowell, the great social and cultural historian, to Franklin Julius Wilson, the great sociologist.

The last word belongs to Frederick Law Olmsted (1822–1903), who was known not only for his talents as a landscape architect but also for his book *The Cotton Kingdom.* When traveling through the antebellum American south, Olmsted found laws against educating blacks defended on the grounds that blacks could no more learn to read or write than could animals or maniacs. He asked, why, then, there were no laws on the books forbidding people to teach animals and maniacs how to read and write.

Nietzsche (1844–1900) admits into his circle of moral concern only the creative geniuses he calls supermen. He commits no logical or scientific blunder in doing that: these choices are based on what each person finds within, and cannot themselves be refuted. Unlike Jainists, for example, most of us do not admit insects to our circle of moral concern.

As usual we must focus on the bridging propositions. For Nietzsche these are:

One, supermen are the sole source of value in the universe. They owe justice to one another but can treat ordinary people, or "herd men", as mere means to their great ends. It is legitimate to murder an ordinary person to keep a great artist in paint and canvas. It is legitimate to sacrifice the lives of thousands of peasants to build a cathedral, or millions of lives to produce a Napoléon.

Two, the masses should be kept illiterate, otherwise they will read newspapers, form political parties, and set up social democratic societies. They will eliminate suffering, and great art cannot flourish without suffering. They will make the herd morality of mindless equality king at the expense of aristocratic ideals of excellence.

Three, a herd society is destructive of great men and great achievement. The emerging superman will be hopelessly corrupted if bathed in mass culture. He must keep his distance. This is really possible only in a caste society, and the best of these have been based on a conquering élite who imposed caste on the vanquished.

Four, the organizing principle of human society should be maximum malleability at the command of supermen. Both Christianity—the debased equalizing enemy—and the idiotic creed of nationalism can be used to manipulate the masses. Look at how Napoléon was made possible by French nationalism, and how many great generals have used priests to bless their troops. The ultimate aim—the sole aim if European civilization is to be salvaged—is a United States of Europe ruled by supermen.

The pill argument and pure logic are ineffective against Nietzsche. If I ask him what he would say if I sneaked into his food a pill that turned him into a herd man, he would say, "Of course I would reject the idea I should be a mere means to the ends of the great. But that is because I am no longer Nietzsche with the traits of a great man but have become a mediocrity. You can change skin color without changing the person, but you cannot change a person into a dog without making him worthy of being treated like a dog. As for logical consistency, if I ever ran into a herd man who saw he should be a mere means to greatness, I would praise him for his perspicacity just as much as I would a superman."

Nietzsche has survived item three but he is vulnerable when we get to items four, five and six. This is because we can use history and social science against him with devastating effect.

Let's start with history. Nietzsche lauds barbarian conquests. However, these conquests did not really impose a genetic or cultural élite on a mass of herd men. Until about AD 1500, when Europeans achieved a technology potent enough to withstand nomadic cavalry, the horse was the greatest instrument of conquest in Eurasia. The only superiority required to be a barbarian conqueror was a homeland with abundant horses and pasture, agriculture that was not sufficiently developed for large permanent settlements, and proximity to a civilization possessing advanced metallurgy. Even non-barbarian conquerors such as the Romans showed no signs of genetic superiority to the Etruscans, Celts or Greeks. I am not taking a dogmatic stance on the possibility there could be some genetic differences between conquering and conquered peoples, but if a gap existed it was light years short of the gap posited between supermen and herd men.

Social science shows that Nietzsche's case for caste is hollow. It freezes in place an élite with no clear superiority. Caste impedes correlation between rank and merit more effectively than any other social experiment that humanity has tried. For even a moderate correlation, twenty to thirty percent of the population must shift class in every generation. An élite established by force and blood relationship must give way to an aristocracy of property or wealth, and finally to an open society stratified by talent free to make its way. The Brahmans of India were as much a dead hand on great achievement as was any other caste. If their semi-monopoly on education and literacy had persisted, much of post-independence India's contribution to the arts, literature, film, science and mathematics would never have occurred.

Only because Nietzsche's ban on education or literacy for the masses has been ignored do we have our own century's explosion of scientific and mathematical achievement. Look at the prospect of a grand unified theory of all the forces of nature, the bold cosmological speculations, the solution of Faltings' theorem, the solution of Fermat's last theorem, the answer to Hilbert's question about Diophantine equations, and the exciting and elegant progress on mathematical curves of genus-2 and above.

To maximize achievements of genius, Nietzsche would have to choose merit with social mobility over caste without merit. This levies demoralizing prices. An open society, one that forces all to compete with some kind of equal opportunity, eliminates the social distance between the élite and the herd that is so dear to Nietzsche's heart. The select man will find that he, and particularly his children, can no longer simply issue commands, avoid the bad company of dwarfed beasts with pretensions to equal rights and demands, and confine the ill-smelling task of studying the many to reading books.

How are the prerogatives of supermen to be transplanted into a socially mobile society? Even bosses cannot use their secretaries as mere means to ends, much less a scientist a lab assistant when the lab assistant may be a scientist tomorrow, or when the scientist's son or daughter is likely to serve an apprenticeship as a lab assistant.

Nietzsche cannot specify any social group likely to become a superman ruling élite. His proposal to breed Jews with the Prussian military to seize control of a united Europe is tongue-in-cheek, a delicious slap at German pretensions and anti-Semitism. Military conquest promises nothing better for the future than it delivered in the past: witness Hitler and the imperial rule of Stalin.

Nietzsche admires the renaissance warrior prince who patronizes the arts. This figure has truly been discarded into the dustbin of history. No general today rides a horse around the field of battle and doubles as a munificent head of state. During Operation Desert Storm, General Colin Powell never got closer to Iraq than Saudi Arabia, and his job as chairman of the joint chiefs of staff was rather like that of a top executive at General Motors. General Norman Schwarzkopf, the commander in the field, played a role akin to someone running a complex computerized dating service operating under pressure. Total automation of reconnaissance and weaponry may soon mean that no "soldier", much less a general, gets within five hundred miles of the enemy until the battle is over.

If Nietzsche can specify no actual or emerging élite that has been staffed by a "select kind of creature", what of the conscious creation of an ideal élite? This poses the problem of identification. It is hard to imagine an institutional method of stamping credentials—a sort of self-perpetuating fraternity-plus-sorority accepting or blackballing candidates—that could operate without self-destructive controversy. After all, the prerogatives of membership include control, enslavement and sacrifice of those rejected.

Nietzsche's own attempts at screening for creative genius do not inspire confidence. He does not list supermen—they belong to the future—but he tells us whom he admires. These include some names we would expect, although Alcibiades and Frederick the Great give pause. He likes Shakespeare, despite the revolting vapors and closeness of the English rabble. Rejected are Bacon, Hobbes, Locke and Hume as unphilosophical, Darwin and Spencer as mediocre intellects, and Schumann because of petty taste. Bach, Mozart, Newton, Leibniz and Gauss go unmentioned. This suggests that no one can identify supermen except idiosyncratically.

Nietzsche can still use his ideals to give advice about personal conduct. Even here the lack of an institutional method of identifying creative geniuses is significant because it leaves open only the alternative of self-identification. Every fool in Greenwich Village who paints him- or herself blue and rolls across a canvas is free to claim the prerogatives of a superman. He or she can murder "ordinary" people in alleys to get money for paint and materials, or even for inspiration. The only real-world consequence of putting Nietzsche's ethics into practise would probably be an increase in New York City's random murder rate. What contribution this would make to great achievement is unclear.

Nietzsche's psychology is not that of an actual living creative genius. Few such geniuses have believed that embracing "everything evil, frightful, tyrannical, brutal and snake-like in man" would enhance their creativity.

Let's now consider the nationalist. He or she is immune to both the pill (item three) and the accusation that they cannot use their ideals to organize a human society (item five). A pill that turned a German into a Frenchman leaves the thesis that Germans are to have preferment intact. Germany is already a functioning human society. A German nationalist's ideals give his nation prerogatives on the international scene; these spring from love of his nation. However, I will show that he is vulnerable to item six: his kind of love is a reality-denying sort of "love" and we can use reason and evidence to demonstrate this.

If a person says they love God, we have a right to ask what it is about God they love (perhaps benevolence) and if they know of an existing entity that deserves the name "God". When Nietzsche said he admired supermen, he told us what he admired about them—creative genius—and at least tried to list some proto-supermen. Someone may tell me he loves unicorns, but I will want to know what he loves about them—perhaps a horn with medicinal powers—and where the unicorns are. The German nationalist must state at least two bridging propositions: what it is about Germany he or she loves, and what evidence they have that the ideal Germany matches the actual Germany.

You can name the individuals you truly love. In the case of a nation, you cannot identify the preferred group by referring to everyone who exists within its borders or holds citizenship: a foreign conqueror may invade your territory, annex it and abolish citizenship. You will have to define the group by those who participate in its culture, speak its language, appreciate its literature and cuisine, or have the personal traits—bravery, honesty, Gallic wit, or what have you—that characterize its people.

We know the language gambit is a lie. No predator nation would have spared its victims if they had all agreed to attend a language school. As for culture, do you mean to exclude those Germans who like Russian novels, Italian opera,

Finnish interior design, French painting and Shakespeare? What of the fact that the mass of your people do not care about any German high culture and prefer escapist literature to Thomas Mann, rap to Bach, McDonalds to local cuisine, American jeans to traditional dress, and Monty Python reruns to any other form of humor?

As for admired personal traits, not the slightest effort is made to verify their presence. The social sciences are not used to discern how many Germans are braver than Turks, how many are more honest than Canadians, or how many are more altruistic than Finns. All attempts to define "superior" national identity have been appeals to fictions such as "blood", race, and a mythic past from which descent is claimed. The Nazis' ideal type never gave reality even a nod, hence the gibe, "Blond like Hitler, slender like Göring, and tall like Goebbels."

This may seem an evasion. Why should the nationalist put forward claims that make him subject to falsification? Why not simply say he loves or admires all those who would identify themselves as German? Well, if he says this he is claiming to love people he has never met and admire people of whose traits he is ignorant.

A woman suspicious of a friend who always flirts with her husband is in touch with the real world, but what if she says she is suspicious of everyone in Chicago and Omaha? A man who says he is angry with his neighbor makes sense, but what if he says he is angry with the Eskimos and the

Amazonians? To say you love your spouse, your children and your grandchildren is rational: these are people you know, with whom you have intimate contact. But to love all Germans—people you have never met and will never meet—is like saying you enjoyed someone else's dinner. To admire a friend's qualities, you must get to know him or her well. Can you say you admire the qualities of all Germans and yet be in complete ignorance of the details of their daily lives? To say such things is to make love or admiration into reality-ignorers. You are like someone who says jokes they have never heard are funny.

To lose touch with reality is the greatest sin against reason. We exempt love only because we revere it so much. We forget that claims about deep feelings are to be taken at face value only if they have real objects, not fictitious or empty ones. To allow the love of real people to blind you to their faults is human. It may be that the human race is constructed so that love will always distort reality to some degree.

Our evolutionary history probably endows us with a capacity to identity with a larger group as if its members were our immediate family. For millennia, we competed for survival with other people who identified with their group in that way.

This kind of love or fellow feeling does not give us a license to kill. If we encountered Martians who were programmed to love unicorns and believe they were real, we might assess the Martians in several ways, but none would include a judgment that they were rational. Rather, we would conclude that their natures forced them to be profoundly irrational.

No modern state treats its citizens as if they are hunting dogs that simply leap at quarry when the horn is sounded. Caricatures of the enemy are posted, showing them with buck teeth (Japanese) or huge noses (Jews) to portray them as inferior. If a traditional enemy becomes an ally, the caricature suddenly turns into a noble-looking creature. During the German–Japanese alliance of 1936, the two states pictured their peoples as if they were morphing into a single physical type. When your nation attacks another, a pretext is invented. No matter how enfeebled the "enemy", the attack is a pre-emptive strike needed to ward off a threat.

The stereotypes and "duties" that your socialization as a patriot has created can be tested against logic and evidence. Made explicit, they are seen to be appeals to cultural icons, superior traits or love of a people, none of which have any connection with reality. Every set of moral principles can be tested as to whether its plausibility rests on sweeping some of what science tells us under the carpet. Those generated by nationalism are no exception.

There has been much talk in my lifetime about the "soil" of one's native land—the holy land of Ireland, the biblical birthright of Israel, the sacred island of Okinawa— as if we were primitive agriculturalists tilling acres with which our ancestors mixed sweat for generations. In 1914, French troops would sneak across the border and bury their faces in the sacred soil of Alsace-Lorraine, which had been annexed by Germany in the Franco–Prussian War. I like to fantasize that the local farmers had brought in topsoil from the next province so the soldiers really tasted the soil of the Rhineland.

If anyone starts listing the particulars of a landscape—perhaps the sublime emotions they feel when contemplating the Bavarian Alps—as something that gives them reason to automatically obey their nation's dictates, give them a photo interspersed with photos of the Southern Alps, the Andes and the Rocky Mountains. Ask them to pick out the one that is so uniquely inspiring.

I hope no one will say the nationalist is logically consistent because he or she gives others the right to be chauvinists for their own nations. First, this is often not so. Second, when it is, the person is merely tolerant of their being as irrational as he or she is. Let them extend their tolerance to the champions of unicorns.

Showing that romantic nationalism is irrational may seem to prove too much. Human societies could not exist without some bonds between their citizens that transcend enlightened self-interest. For a rational person these unifying bonds must be based on moral ties, such as a communal pursuit of justice, rather than on personalized emotion: I do not love my fellow New Zealanders indiscriminately, or exaggerate their individual virtues. However, national societies are, for now, the largest groupings whose people can cooperate to build humane societies. There is nothing wrong with enjoying your national or local or ethnic culture. Be proud, so long as the culture nourishes the good.

Everyone needs citizenship to survive, and self-preservation is morally permissible. Today's rational person is adrift in a world organized into the equivalent of mobs of soccer hoodlums, whose passions blind them to the feelings of other peoples and make them capable of mindless violence. You must belong to a mob for protection against other mobs. If your nation is attacked and you want others to risk their lives for your defense, you must risk yours. If your nation is in the right, as in the war against Hitler, you will fight with moral passion. If your nation is doing a great wrong, be a conscientious objector if that is allowed. If you must serve, discharge your gun into the air except when the safety of you or your comrades is at stake.

We owe it to ourselves to imagine every difficulty the agenda poses for humane ideals, because we should want to know if our ideals are flawed. The most devastating possibility we face is that our ideals could self-destruct in practice. This is discussed in the next chapter. As for the points raised so far, we must be tough-minded enough to face up to hard choices.

Psychology has shown that chimpanzees have self-awareness and a sense of justice. It is hard to see that it is allowable to use them in medical experiments unless we would be equally willing to use a human. Indeed, they are less eligible in that a human might volunteer.

If economics shows that infrastructure is crucial to eliminating poverty, the welfare state may have to make do with minimal funds until some degree of affluence is achieved.

If you cannot include everyone in your circle of moral concern because of gender, or membership in the wrong class, you should be honest enough to admit that your humanism is qualified by a privileged position for part of humanity. You join the ranks of our opponents, and I predict you will have difficulty squaring your hierarchy with the items on our agenda. No good argument can be given as to why a male cannot be committed to justice for women, or why writing off a class because of its social indoctrination would not apply equally to a race or ethnic group, or at least to many individuals within those groups.

Unless we build sound bridges between our humane proclivities and the real world, we cannot make the world into something better. Reality always defeats illusions, even illusions that appeal to nice people.

What of racists who will not look at the evidence, or who look at it and still hate black skin? If a flat-Earther will not look at the evidence, he cannot be convinced that the world is round. I cannot convince a stone of anything. It does not discredit philosophy that reason is effective only with the rational. It is philosophy's job to tell me whether I can hold humane ideals and be a rational agent, and whether another person can hold anti-humane ideals and be a rational agent.

You need not win arguments with ignorant others: the debate that counts is the ideal debate you conduct inside your own mind. You can make sure that its parties scrupulously observe the rules of logic and evidence.

There are tactics other than debate that may be effective in an irrational world. When Martin Luther King (1929–1968) went to Montgomery, Alabama, most whites were impervious to evidence that blacks were not permanent children. When they saw that blacks had the self-control to boycott the buses, the intelligence to operate a complex car pool, and the courage to accept violence without retaliation, some whites began to change their minds. They said to one another, "I am not sure I could undergo all of that for what I believe." Highly visible evidence that stereotypes are misleading can be more effective than either reason or rhetoric.

If there is nothing outside ourselves that can claim the right to tell us what our ideals should be, our ideals can neither pass (be objective) nor fail (be subjective). They are worth exactly what they are worth to us.

That is just the beginning, of course. The next step is to test them against our agenda.

When people look within, some find humane moral commitments; others find commitments to race, tribe or supermen. To have a developed ethics, rather than mere proclivities, you have to bridge the gap between those commitments and the real world. The bridging propositions of humane egalitarians can be rendered logically consistent and scientifically respectable; the bridging propositions of the anti-humane cannot. Nonetheless, the brute fact remains: we cannot convert anyone to humane-egalitarian ideals without changing the inner person.

I was once emotionally wedded to the notion that an external "authority" could replace the inner person as the source of moral ideals. Now I take satisfaction in the fact that my duties are self-imposed. It confers a certain dignity. This was encapsulated for me by Thornton Wilder, the great American novelist. "How terrifying and glorious the role of man," he wrote, "if, indeed, without guidance and without consultation he must create from his own vitals the meaning for his existence and write the rules whereby he lives."

What is Possible?

IS THE GOOD SOCIETY POSSIBLE?

In the last chapter I argued that Nietzsche would find it impossible to use his ideals to organize a human society. In 1973 I read Richard J. Herrnstein, the prominent Harvard psychologist. Herrnstein argued that those of us who hold humane egalitarian ideals would find it impossible to organize a human society. We would have to either rank people by caste, which would be unjust but leave some people of merit at all levels of society, or rank people by merit, which would be just but produce a lower caste composed of the dregs of humanity, a sort of genetic dump. In other words, the good society was impossible. It took me twenty-five years to find a rebuttal.

First, some terminology to distinguish political philosophy from related disciplines. The study of political behavior is applied social science. The study of international relations is applied mass psychology: it uses concepts such as national interest, which varies with how powerful you are, national self-image, and national affinities—positive or negative profiles of other nations—to explain how states and other international actors behave. Political philosophy, which is what we are about to do here, links philosophy proper and political studies. It is applied ethics: its focus is less on how individuals should behave and more on how we should organize ourselves into a civil society.

Those who believe in humane ideals will want a civil society that counts every person as worthy of moral concern, which is the essence of civic virtue, that emphasizes the six great goods, and that has developed trade-offs between these goods that people find acceptable.

The six great goods are by now familiar: human happiness, justice, the pursuit of truth, the creation of beauty, the perfection of the individual, and tolerance of diversity. Based on history—whose lessons can, of course, be debated—these great goods imply certain organizing principles. The welfare state reduces human misery and ensures people's lives are not too much blighted by bad luck (which would be unjust). The pursuit of truth and the creation of beauty flourish where there is democracy, free speech and artistic freedom.

The perfection of the individual requires that each person's capacity to attain his or her own excellence is not too circumscribed by what other people are willing to pay them to do—in other words by the "market". This requires a reasonable degree of affluence. It also requires that society's mores don't blind people to the truth about human excellence, which happens when people are corrupted by militarism or materialism. Tolerance of diversity requires the absence of racism, sexism, ethnic chauvinism and stultifying orthodoxy, both religious and secular.

If justice requires that people neither profit nor suffer unduly because of luck, this includes the luck of birth. Therefore, two important organizing principles of a humane society are the reduction of privilege and the reduction of environmental inequality. The meritocracy thesis argues—based on knowledge of how genetic differences affect people's achievements and behavior—that these two organizing principles self-destruct in practice.

Herrnstein's thesis says: Assume we make progress towards the equalization of environments. To the degree that this occurs, all remaining differences in peoples' talents will be due to differences in genes. Assume, also, that we make progress towards the abolition of privilege so that only talent counts. To the degree this occurs, there will be a social mobility that brings all good genes to the top and allows all bad genes to sink to the bottom. The upper classes will now become a genetic élite whose children inherit their status because of superior merit, while the lower classes become a self-perpetuating genetic dump, too stupid to be of use in the modern world, an underclass that is underemployed, criminal, and prone to drugs and illegitimacy.

It is certainly true that genes influence many human traits—IQ is an example. I would be the last person to equate human merit with IQ but there is no doubt that there are IQ thresholds for various occupations in most countries. Few below an IQ of 110 will qualify for élite professions such as medicine, law or science. Few below the average IQ of 100 will fill managerial, technical or professional occupations of any sort.

Therefore, the meritocracy thesis generates a prediction. If the children of the upper classes, thanks to better and better genes, are becoming more eligible for high status, and the children of the lower classes, thanks to worse and worse genes, are becoming less eligible for anything but low status, we should detect a trend. The class IQ gap—that is, the gap between the children of the top third in occupational status and the children of the bottom third in occupational status—should widen over time.

I tested this hypothesis against evidence for United States children aged six to sixteen and found no such trend. The class IQ gap had been remarkably stable from 1932 to 1989; indeed, it may have diminished slightly from 12 IQ points to 10 or 11.

The mean IQ of the children of the lower third of parents is steady at about 95. I will chance my arm and predict that it will not sink below 93, which would leave the children of the middle third of parents at 100 and the children of the top third at 107. This applies only to white America. Thanks to lack of opportunity, what IQ a black American had once counted for very little. Today, black Americans are more stratified for IQ than in the heyday of racial bias.

What would the kind of IQ gap I found among white children mean in terms of social mobility? Children of middle socio-economic status would be split evenly, with 50 percent in the top half of the IQ scale and 50 percent in the bottom half. Children of low socio-economic status would be 32 percent in the top half of the IQ scale and 68 percent in the bottom half. Children of high socio-economic status would be 68 percent in the top half and 32 percent in the bottom half. To the extent to which IQ counts in social mobility, this implies a high rate of children of all classes shuttling about the occupational scale generation after generation. It hardly conveys a picture of an American underclass permanently unemployed, criminal, and prone to drugs and illegitimacy.

However, this kind of evidential refutation of the meritocracy thesis leaves the core of the thesis untouched. The obvious rebuttal is: Well, if the genes/merit gap between the upper and lower classes has not expanded, this merely shows there has been no trend towards the erosion of privilege and environmental inequality. The only reason your egalitarian ideals have not produced an anti-egalitarian nightmare is because all of your efforts to promote social equality have failed. You must pray for eternal failure. If you ever did succeed in putting your organizing principles to work, your ideals would self-destruct.

The fact remains that if the meritocracy thesis were true, the good society would be impossible. Whatever ideal the thousands who laid down their lives for social reform or the defense of the republic may have had, it was not a class system frozen into a caste system by a genetic inequality enhanced by every step towards social justice. Moral debate would have dealt humane ideals a crushing blow (see item four: Show that your ideals can organize a human society). Just as Nietzsche could not use his ideals to organize an aristocratic society, we could not use humane ideals to organize an egalitarian society.

Let us imagine the good society as a going concern and see if it would degenerate into a meritocracy. After all, if a society's organizing principles self-destruct in practice, the society should prove unstable.

Imagine a residential college or university where everyone is fully funded and has forgotten that one day they must earn a living. Everyone takes classes, but there is plenty of free time to do whatever you most enjoy, or feel will most perfect your talents. Some play chess, some join the school literary magazine, some the theatre, some play sport, some like to work with their hands at craft or construction, some like to socialize at the local pub, and some, of course, do a multiplicity of things. Everyone has a reasonable amount of pocket money to enjoy a social life.

There is no privilege, or inequality of environment. Everyone has access to the tools and training they need, and insofar as there is a hierarchy within activities it is created purely by excellence of performance. However, the students are not obsessed with hierarchy to the point of lacking satisfaction in doing the best they can. Someone who, after training and trying, runs a four-and-a-half-minute mile is exhilarated, even though he cannot run a four-minute mile. Someone who writes poems that please her does not brood too much over the prolific talent of her neighbor. Someone who designs and builds a seaworthy boat does not have to be the best carpenter at the college. Someone who attracts a good partner does not lament that their partner is not the most good-looking or charismatic person around.

These students mate and have children. Would they breed themselves like racehorses, with the best male and female chess player pairing off to produce the champions of the future? Not if they have the normal delight in diversity. The best male chess player may well fall in love with the most charming girl he meets at the pub. The best actress may want a male athlete. So it is most unlikely that for each excellence, generation by generation there will emerge a more and more radical genetic hierarchy. The best female and male chess player may get a surprise when their child, despite his or her wonderful chess genes, goes for rock music. Many, of course, are not even aware of their best talent.

What could cause this good society to degenerate? After four years the citizens get a shock. To stay in the community, they must now pay their own way by earning a living. Worse, they find themselves in a poverty economy. In order to survive they must work with heart and soul at whatever pays the best. Excellences no longer count unless they are valued by the market.

A few people are lucky: they are so good at acting or chess or carpentry or sport that they are paid to do what they love anyway. Some find jobs they did not anticipate they would love—perhaps the drama of being a criminal lawyer—but most have to do something that derails them from their special interest or excellence. They have to devote themselves completely to the entrepreneurial virtues, to winning the race to get as much money as they can.

So now they are in something like an annual school race on which everything depends. Such a race, if everyone has access to equal diet and training and everyone single-mindedly gives their best, will rank them in a hierarchy for running genes. And if nothing else is valued, like may mate with like.

However, will this poverty economy really abolish privilege and equalize environments? None ever has. In subsistence economies everyone seizes on whatever scrap of privilege they can, even if it's only having a relative who is a minor civil servant. And the few rich people, who could finance a minimal welfare state in order to make the environments of the poor a bit less crippling, are not usually so inclined.

Now let us assume that after four years at the residential college, the students must make their way in an economy of affluence. Once again a few are lucky: they happen to be very good at something that the market rewards, or they discover in their job something they love. They can go on much as before. But need the rest deviate from their accustomed patterns of behavior? Only if they are infected by materialism. If moderate effort at a job, plus a good welfare state to provide against misfortune, gives them all the money and security they need, they can go on with the multiplicity of good lives they have always enjoyed.

So the good society degenerates not because of the abolition of privilege or inequality of environment, but because of poverty and the single-minded struggle to survive it engenders—or because, even under conditions of affluence, its people are corrupted by materialism and run the money race with near total dedication.

Let us probe the psychology of a people who do become obsessed with the competition for wealth, devote all their time to it, mate for success in it, and create something of a hierarchy for whatever genes influence the traits that make for victory. If they are to keep the race fair, these money-obsessed people must give the children of the losers a very good welfare state to ensure they can compete on an equal footing. That means large transfer payments—progressive taxes—out of their pockets so other people's children will have a better chance to beat their own.

It also means that money obsession must be strong enough to override the fact that everyone can, thanks to the welfare state, have a decent life even if they do not do well in the competition. For many, this is likely to weaken the incentive to devote themselves single-mindedly to the race: certainly they do not have to amass the capital to guard against misfortune. If that security is combined with parental affluence, it may be "corrupting". I know of many a professional parent who is supplementing the welfare state to support a son who earns little from his art, or a daughter who earns little for her acting.

Does the meritocracy thesis have a coherent psychological foundation? People must be both money-drunk and justice-drunk. This is a rare combination. And their money lust must not be too diminished by the fact that a decent life is widely available even if you are not dedicated to market competition.

I return to my finding that America has not shown a trend to meritocracy. Other recent studies show the same. America seems static, and in England the association between IQ and occupational status has, if anything, been declining.

Now, we can put an optimistic spin on the absence of a trend. It may not be due to any failure to abolish privilege or environmental inequality. It may be due to the fact that as the people in developed nations have become accustomed to affluence, which may take a few generations, the obsession with the money race has declined. Perhaps the single-minded devotion to the competition for occupational status and wealth, which is necessary to get a high correlation between class and genes for IQ, is on the wane.

More and more people may be following their own star— building thousands of hierarchies based on diverse merits, only loosely correlated with one another, not necessarily even following their best talent (I enjoy running more than mathematics), mating with people of different interests and talents from their own, and having children whose choice of what genetic hierarchy they want to ascend is unpredictable. The French have chosen to work thirty-five hours a week rather than strain to boost productivity. Polls show that the average Finn would not trade a job they enjoy for one less satisfying with double the pay. People may be acting as if they were human.

What would be a "perfect" meritocracy of the sort that Herrnstein describes? It will be a competition that best realizes a hierarchy of merit. The prerequisites are that it selects for a single trait, under conditions of absolute fairness, and the incentive system is such that all try as hard as they can. Under these conditions, if the trait is heavily genetically influenced the competition will create a genetic hierarchy. If like mates with like, that hierarchy will be transmitted from generation to generation to the maximum degree.

Imagine that at another university income is totally dependent on the annual school race: those who do best get ample funds for food and lodging and pleasure and those who do worst get nothing. During the year that they train for the race everyone is well fed and housed and gets the best coaching available. No one gets a head start. The incentives to train and to try on the day are absolute, and there is maximum fairness. There is little doubt that performance would sort the student body by genes for distance running. We will imagine that reproduction is by cloning and that every year each student produces an infant dependent on its "parent" for rearing and sustenance. These represent the next generation of runners.

What would cause this meritocracy to degenerate? If those who do worst at the end of the year are left to eke out an existence on charity, their offspring will suffer the effects of poverty, such as poor diet and no medical care. To avoid this, the prizes are altered: the winners still get affluence but the losers get everything they need except money for pleasure. Indeed, you can get a bit of money, not much but some, from doing what you like best: chess, algebra, the school paper, the poetry society and other sports, all of the things that are alternatives to distance running.

This alters the incentive system. Few will now do the full Lydiard training schedule of running 100 miles a week. Most will settle for the 15 miles a week that is sufficient training for them to race at ten seconds per mile slower than their optimum pace. Moreover, every individual who does this lowers the quality of performance needed to run an average time in the race. Once most people are running only 15 miles a week, if you have talent that is a bit above average you can do pretty well by training only ten miles a week. This will further lower the average performance, which will further lower the training you need to do to be average, in a downward spiral. The school race has degenerated in the sense that it no longer ranks people very well, even for their running genes.

There is an incongruity that the advocates of the meritocracy thesis never face. As it is supposed to be a serious hypothesis about the effects of egalitarian policies, you would expect it to be grounded in evidence. When I read it, I expected it to be supplemented by a list of all advanced nations, ranked in terms of how far they had carried the equalization of environments, and those same nations ranked by the size of their demoralized underclass. As everyone knows, these two hierarchies would correlate not positively but negatively: the Scandinavian nations—Sweden, Norway, Finland—would stand near the top in terms of their welfare states, but near the bottom in terms of the percentage of their demoralized citizens.

Does the meritocracy thesis have a coherent sociological foundation? To sort for genes it must be fair. To be fair it assumes a robust welfare state. The welfare state both "causes" and "prevents" a meritocracy at the same time.

I cannot emphasize too strongly that a robust welfare state is not a gratuitous boon: it is the very soul of a true meritocracy. The notion that a meritocracy of any sort could lead to an underclass is absurd, unless the "meritocracy" is to be a shooting star that persists for one generation. If environments are to be even roughly equal, and the sins of the parents are not to blight the lives of the children, the lack of merit of the parents must be ignored to the degree that is necessary to provide every child with a non-demoralized home, good diet, good health and good education. If there is even a tendency towards an underclass, their environments must be topped up. In other words, to sort efficiently for genes the consequences of doing badly must be draconian, and yet the consequences cannot be draconian.

The incoherence of the meritocracy thesis can be summarized as follows:

Psychological incoherence. First, meritocracy assumes that people are motivated to run the money race by love of money; second, to ensure fairness those same people must be willing to transfer huge sums of money out of their own pockets to others so environmental inequality will be moderated.

Sociological incoherence. First, meritocracy assumes rewards will be allocated according to merit; second, to ensure fairness rewards must be allocated to a large degree irrespective of merit (the welfare state).

Let us concede, for the sake of argument, that people in a particular society can be both money-drunk and justice-drunk: despite their desire to maximize personal wealth, they are committed to the alleviation of privilege and environmental inequality. Therefore, a true meritocracy can exist.

However, it is noteworthy that the same things cause both the true meritocracy and the good society to degenerate. No matter whether incentives are focused on the competition for money or on all the endeavors that make for self-realization, privilege and environmental inequality must not intrude. Otherwise, neither one competition nor many competitions will be fair. This means that neither meritocracy nor the good society has an advantage in terms of whether its organizing principles self-destruct in practice.

Their commonality comes from the fact they both incorporate the great good justice. Well, better that "ball" than none at all, but equal opportunity leaves open the question: equal opportunity to do what? R. H. Tawney (1880–1962) offers a brilliant image: if an elephant is let loose in a crowd, everyone, except the beast and its rider, has an equal chance of being trampled. It is not much good to live in a society that gives everyone a fair chance of being a parody of a human being (see box on the next page).

The humane society will incorporate all six great goods. Now we see that the good society pays no price whatsoever for this in terms of incoherence. If anything is unstable it is meritocracy, because of the unlikely psychology it requires. Neither society will tend to evolve into the other unless there is a sea change in mores. Poverty is a common enemy in that it both erodes altruism and renders the pursuit of excellence impossible by most.

There is one way in which the two societies might coincide: a "magic market" might happen to reward all of the excellences any sane person wants to pursue. Who would object to that? It would be identical to our good-society residential university before graduation.

Since no market has ever done this, the state should try to simulate one. It should, using public funds, try to soften the conflict between a person doing what they wish to do and doing what others are willing to pay them to do. Assuming affluence and much leisure, the state can provide facilities for art, crafts and sports, so that everyone who wishes can

TAWNEY AND THE GOOD SOCIETY

Tawney says that if a person has the opportunity to perfect his or her talents, and enough money to do that properly, they have all the happiness that is good for any of the children of Adam.

Herrnstein, along with Charles Murray, later spelled out the meritocracy thesis in a book called The Bell Curve (1994). He presents no image of what America would look like if it were a good society, except that we would prevent the average IQ from diminishing, with attendant benefits such as less crime and a more useful work force.

Unless the average IQ went very high indeed, the hopeless underclass would diminish but still remain sizable. What hope is offered to this class? Herrnstein and Murray say that everyone can hope to be valued by his or her loved ones, despite the fact he or she has no work to do and is criminal, violent and drug-ridden. We encounter the same poverty of ideals we found in William James and economists: once you can fulfill your demands by successful participation in a capitalist society, little more need be said. No doubt Herrnstein and Murray hoped the successful would spend their leisure in civilized pursuits, but that requires a full-fledged assessment of the values and incentive systems of contemporary America. Civilized people grow in civilized soil.

have ready access to the venues, equipment, training and tools they need; subsidized semi-professional baseball and subsidized workshops can join local theatre and music. The local pub is usually self-financing.

Aristotle thought of Athens and Carthage as two relatively affluent societies that had "chosen" different roads of development. Neither provided a welfare state, but they had solved the problem of generational poverty through emigration. The unsuccessful, the landless, went abroad to found new city states. Athens held up the ideal of leisure devoted to civic virtue and the six great goods, except for delight in diversity; Carthage identified the ideal person with the ideal entrepreneur.

Athens produced art, philosophy, geometry and wonderful artifacts, and had a citizen's army. Carthage was a commercial society—Kipling called it a sort of god-forsaken African Manchester—in which people were socialized not to want to do anything the market did not reward. Therefore, they created no great art or theatre but were content to be consumers and spectators, happy to buy paintings and purchase tickets to performances. For defense they hired mercenaries. Their navy once went on strike for higher wages when faced by an enemy fleet bent on invasion.

These two do not exhaust the options. In theory a society could choose to cultivate military virtues and live by plunder. If all could ascend to the rank of general by merit, it would be a meritocracy staffed by genes for the military virtues.

It is easy to criticize the meritocracy thesis. It assumes that materialism will not affect the willingness of the rich to make the transfer payments needed to equalize environments. It assumes that equal opportunity is compatible with the existence of an underclass, as if equalized environments need exist for only one generation. It provides no evidence that nations which have done the most to equalize environments have the largest underclass. Whatever demons lurk in the depths of equality, meritocracy is not among them

It is not easy to actually realize the good society. It is not enough to have democracy and freedom of speech, and eschew militarism and imperialism. The good society has an eternal and powerful adversary: materialism.

Affluence sets a test every society must pass. It must choose between Athens and Carthage. Every factor that influences human history influences what is chosen, but the individual is not impotent. Everyone above the poverty line in a developed nation can make a personal choice between materialism and humanism. You can choose to be a corporate lawyer who works ninety hours a week and evades taxes, or select the occupation that least hampers you as you walk your own road to personal excellence and promotion of the common good. When any individual chooses the latter, the prevailing materialism is by that much diluted. When anyone chooses the former, the good society recedes. Politics can never sever its ties with ethics.

IS FREE WILL POSSIBLE?

We now go from what is sociologically possible to what is psychologically possible. Everyone at some stage of their life has wondered whether human beings make at least some free choices, or whether all our choices are determined by forces over which we have no control. What is at stake is whether we can or should levy moral praise or blame on a person for their behavior.

Is the present self really free when it makes morally significant choices (and many trivial ones as well)? By the "present self" I mean you at this moment deciding between visiting a sick friend or going to an escapist film. You entertain the appearance of freedom—that is, you must choose as if you were free. You cannot just sit and wait for a billiard ball to knock you towards the hospital or the movie theater.

The question is whether all such appearances are deceptive because an underlying reality, perhaps our brain physiology, determines all our decisions. Most people (but not most philosophers) believe that if our behavior is so determined we cannot be praised or blamed for our decisions. We do not blame a clock for striking the hour ten minutes early. Even if the clock had a conscious life and the illusion of free choice I would not blame it.

Ever since Gilbert Ryle (1900–1976) coined the phrase, those who believe people make choices that are not determined have been accused of believing in a "ghost in a machine". Supposedly these people turn the conscious mind into a mysterious thing that lives in the brain but has no connection with it, except to give it orders.

This is not the case as far as I am concerned. I believe the human mind is a functional system with both unconscious components—whatever is going on in the brain—and conscious components— whatever you are thinking about at this moment. Sometimes my brain is active when I am not aware of it: whole pages of a book break through to consciousness already composed. Sometimes what is going on in my consciousness—anger, for example—has effects my body cannot control, such as stomach pains. But the concept of free choice does mean this: my brain can cause states of mind that have the peculiar attribute of autonomy.

I see nothing odd in this because the stuff from which the physical universe is made has sprung many surprises as it has evolved. At one time it was too hot and active to allow for matter in the form of things made out of compounds. Then along came long-chain carbon compounds that were self-replicating or alive. Next, some of the organisms composed of the right compounds achieved consciousness. Then some of these achieved a sense of personal identity. Now, perhaps, some have achieved a conscious personal identity that can, to some degree, make choices without interference from anything else. This possibility can be discounted only if the concept of a free present self is logically incoherent.

Here we must watch our step. At every stage a case for incoherence could have been made. How can matter that is "dead" possibly spawn something living? How can matter that is unaware spawn something that is aware? How can matter that has no "self" spawn something that has self-awareness? How can something that obeys laws spawn something that is free? The possibilities of matter cannot be circumscribed by logic.

The concept of free choice is perfectly coherent and easily stated. Free choice, to the extent that it is real, would be a caused first cause. The state of mind in which I make free choices is an effect, of course. My awareness of choices began at a certain age and will disappear when I die. Once this awareness comes about it has the power to alter the flow of the world from past to future. If free choice exists, the present self has a genuine choice between (at least two) alternatives, and creates a future that would not otherwise have existed. If all of us, at increased risk to our lives, decide to pick up hitchhikers as an act of charity, the world will be different: more hitchhikers will get to their destinations quicker and some extra lives will be lost.

The Jesuit psychologist Father Michael Maher (1860–1918) said, "Besides the motives felt, and besides the formed habits or past self, is there not a *present self* that has a part to perform in reference to them both? Is there not a causal self, over and above the caused self (the character) that has been left as a deposit from previous behavior?"

As this implies, free choice assumes that the will is to some degree "self-generating". Along with many other things, the free choices of the past affect the performance of the present self. The more good choices I make, the more I enhance "willpower"—that is, the more the present self will find it easier to choose good over evil. Choices affect character. When I act out of regard for moral principles, I enliven my commitment to them.

The relationship between the present self and character is important. It is sometimes said that if my present self is free it must bear no relation to my character: it must be some kind of abstraction that has nothing to do with me. And, if so, it is just some kind of loose cannon acting in a void, and so how can we give it moral credit for what it does?

The answer is that my present self is a rather important part of me, and when I give "it" credit I give myself credit. My present self has been my faithful companion throughout life, the part of me that has had to make choices, the part of me that has recorded a history with (at least some) good choices, for each of which it deserves credit, the part of me that deserves some credit for my virtuous (on balance, I hope) character today, although only insofar as it has played a role in the evolution of my character. I deserve no credit for influences that molded my character if they were beyond my control.

I should add that I have distinguished the present self from character—in the sense of the repository of principles—simply to emphasize its unique role. If you wish, it is that part of my character that must make free choices and record a history of good choices and thereby strengthen my will.

Is free will worth having? Daniel Dennett (born 1942) argues against the dignity of free will on the grounds that it is irrelevant to what we admire most: someone who always does good. However, we do not admire a *thing* that always does good. Unless abused by way of an overdose, aspirin is an almost perfect medication. It does not inspire admiration because it deserves no credit for its effects. I admire people who find it easy to be good, but only because they deserve some credit for what they have become. Their present selves, over time, made a whole series of choices rightly, and some were very difficult. As a result they gained the strength of will to do what moral principles, more and more deeply ingrained, entail.

Thanks to the present self, these decisions are now virtually automatic. Note the word "virtually". Even the saint does not attain the perfect or holy will that Kant attributed to God, and it is hubris for anyone to believe that, like God, he or she is beyond temptation. Those who think of themselves like that are likely to find themselves suddenly at risk—say in old age, as the prospect of death engenders a sense of indifference.

The fact the road to sanctity is paved with free choices is crucial. We take this into account when we give the highest praise to those who have had the most difficult path. For some who are raised humanely with few temptations, the road is not easy, because it is never easy, but we most admire those who become outstandingly good despite adversity.

Does that lead to the odd conclusion that we should not create a good society because virtue would come too easily? Of course not. The fact that moral praiseworthiness is a great good does not mean it is the only great good. If moral praiseworthiness is diminished by a social dynamic that makes humane actions more frequent, then the trade-off is worthwhile. Would we want to create a society in which sanctity was a certain outcome for everyone? That amounts to wishing we were angels rather than human beings, which is as absurd as wishing we were social insects. Human nature is the foundation of the value of moral praiseworthiness, and if you abolish our humanity moral praiseworthiness, of course, loses its raison d'être. Given what we are, we need all the help we can get to become good.

Many readers will think the central question is whether our experience of being free when we make decisions is mere illusion or sometimes corresponds to reality. They will add that we can be praised and blamed only if we are really free. I agree: reality trumps appearance, and if the reality is that all our decisions are causally determined, then moral praise and blame are irrelevant. However, most current philosophers deny this and call themselves "compatibilists": they think they can both explain all human behavior scientifically and still believe in free will, or at least a will deserving of praise or blame.

The compatibilists and I have some common ground. We agree that clocks differ from people. Clocks are unconscious while people are aware of certain thought processes when they make decisions—that is, they entertain considerations, weigh them, know that nothing will happen unless they make a decision, and so forth.

Where we part company is whether meaningful freedom sets limits on scientific explanation. If I return a borrowed book to a friend, the universe is such that he can read it that night. If I do not return it, the universe is different in that he cannot. Hence a free choice creates a radical discontinuity from one state of the universe to the next. Indeed, this is true of every free choice that people make throughout the world. Much human behavior and its effects escape causality in the radical sense that they escape any scientific explanation.

Some compatibilists think qualifying the nature of causality evades the consequence that scientific explanation must be circumscribed if we are to be worthy of praise or blame. They appeal to an argument from David Hume, one that will preoccupy us in a later chapter.

Hume (1711–1776) showed we have no reason to assume necessary connections between events. Compatibilists construe this to mean that things such as decisions can be "caused" without being "necessitated" or determined. In fact, Hume did not think that any kind of causality could survive the demise of necessary connection.

The essence of Hume's argument is that there may be utterly spontaneous occurrences, events that neither science nor anything else can anticipate. If these can occur, and he sees no reason they shouldn't, all causality goes. There is no exemption for some strange "non-determinating" causality.

If we take Hume's argument seriously, we are left with nothing at all: science, bereft of the very concept of causality, can explain nothing.

If Hume's analysis applies to all causal explanations—those that explain the behavior of a clock as much as those that explain the behavior of people—it can hardly be used to differentiate between clocks and people. And yet that is what we must do if we are to justify blaming people in a way in which we do not blame clocks—we must show that alternatives are truly open for people in a way in which they are not open for clocks.

It is true that people make conscious decisions and clocks do not, but what does this have to do with the nature of causal explanation? A pre-scientific person who believes things have in-dwelling spirits may think clocks decide when to strike.

Of course they would be mistaken: the fact that we experience freedom establishes nothing. Many of our experiences are deceptive. Mirages deceive us about where things are. Hallucinations deceive us about what things exist. Who is to say that our experience of decision-making does not deceive us when it implies that more than one alternative is open? Virtually every physiologist I know thinks that these experiences are deceptive, and that science will eventually be able to account for the outcomes of all our decisions from brain states. Does anyone believe the antidote is to recommend reading Hume? How would that obviate anticipating the outcomes of all of our decisions from brain states?

Other thinkers draw a distinction between causes acting from without and those acting from within. Causes acting from without, such as gravity, produce behavior for which I cannot be blamed—for example, I cannot be blamed for falling on someone after being pushed out of a window. However, when forces acting from within the psyche cause human choices, the choices are held to be subject to praise or blame.

The distinction is real but beside the point. A stone is at the mercy of causes outside itself. If someone hits me with a stone, it is silly to call the stone in question a bad stone. It is no different from any other stone, including "innocent" stones. When a clock acts in response to forces from within and deceives me about the time, it makes sense to call it a bad clock: it *is* worse than good clocks. But this is merely a condemnation of its bad nature or "character". It is not moral praise or blame. We could blame it for the character it has only if it had free choices that influenced its character.

Another causal argument appeals to indeterminacy, either on the subatomic level—electrons unpredictably jump from one place to another—or in the context of chaos theory, where thousands of chaotic trends make the timing and occurrence of an event only probable. But indeterminacy takes us no farther than Hume. The electrons of a clock are just as unpredictable in their jumps as those of a human being. Chaos theory applies to things such as predicting the weather, where thousands of chaotic variables come into play, but no one uses it to give a causal analysis of either clocks or the behavior of an individual human being. Like Hume, indeterminacy does not differentiate between clocks and people.

Kant (1724–1804) showed how irrelevant indeterminacy is to the central question of whether we have choices that can be subject to moral praise and blame. He discusses positive and negative freedom. A random event such as an electron jump has only negative freedom. You don't praise or blame a Mexican jumping bean for being unpredictable: it has no more positive freedom than a clock. You can be praised or blamed only if a present self exists and can freely choose between open alternatives.

Some have tried to say Kant was a compatibilist, but in fact he called compatibilism "a wretched subterfuge" and condemned any attempt to reconcile causality in accordance with natural laws with the concept of freedom.

Compatibilists assert that if we posit an alternative to determinism we must believe free choices just pop into existence, which is hardly consistent with our being in control of them. The reader now knows this is not the case. The present self does not pop into existence: it has many causal antecedents (including its own choices). Its choices do not pop into existence: the present self makes them.

Dennett stands out from most compatibilists (see also John Gribbin, born 1946). He concedes that scientific, or naturalist, explanations of how the world works leave no room for radical free will—that is, for a reality that includes free choices as breaking the continuity of the universe from one state to the next. Indeed, he has a field day demolishing those who try to fit free choice into some niche (such as indeterminacy) created by scientific explanation. However, he then rejects the reality of radical free will because it cannot be reconciled with scientific explanation.

Why assume we must reject the reality of free choice if that renders part of reality beyond scientific explanation? Why not assume the reverse, that there are limitations on science if uncaused outcomes of choices are part of reality? Science excites our admiration because of the wonderful explanations it has given us about the world thus far, but no one has provided evidence for the hypothesis that all of reality is susceptible to scientific explanation. As we shall see, science itself may suggest the opposite is the case. Before judging that to be absurd, wait for a discussion of how free will versus determinism might be decided by evidence—at least in theory.

Dennett analyzes the flip of a coin. He rightly notes that the practical significance of this event is not illuminated by the fact that the outcome is causally determined. The speed and vector of the spin, the density of the air and the effects of gravity determine whether the coin falls as heads or tails, but there is no pattern that we can predict. That is the whole point of using a coin as a device for making random choices. We see it as a fair way of giving two alternatives an even chance of selection. Isn't this a case where an undetermined "appearance" trumps a deterministic "reality"?

Let us imagine that the coin was an agent and we leveled an indictment of moral irresponsibility. The two alternatives were whether to visit our sick friend or go to an escapist film. We say to the coin, "You have chosen to make this decision in a way that ignores the significance of the two alternatives. One act is dictated by a moral principle, the other by the pleasure of the moment. What you have done is not as bad as simply giving into the temptation to enjoy oneself at the expense of a friend. Even so, to make the outcome a matter of chance was totally irresponsible."

The coin replies, "But I had no control over the situation. It is true that I have to bear causal responsibility for this decision in the sense that I was a necessary participant in the events that led to it. But others manufactured my character—neither heads nor tails is the heavier side—and once I was in motion forces determined the outcome. You can pass judgment on my behavior as bad but you cannot pass a moral judgment on me for behaving badly."

The coin has correctly emphasized that, as always, reality trumps appearance. Whether we describe its behavior as random or determined, the reality is this: the coin lacks a present self that plays the role of a cause with at least two possible alternatives.

This suggests a distinction. Even when agents are unable to choose other than they do, we can label their character as productive of bad acts—for example, if the coin's decision was against visiting our sick friend. We can force such agents to assume ownership of their acts (the coin admitted that its "character" made it causally responsible for its acts). And we can do something to ensure they do not replicate their conduct (lock up the coin). However, we cannot assign guilt in the sense of calling them blameworthy.

If scientific explanation extends to the whole of reality—including all the choices of my present self—free choice is mere appearance. And if that is so you can judge my behavior but not me. I will give these two kinds of judgments different labels. Assessing good or bad behavior is giving moral approval or condemnation. Assessing the culpability of the agent is allocating moral praise or blame.

I have maintained that unless people are free to choose between one thing and another they are not morally praiseworthy. The American philosopher Harry Frankfurt (born 1929) questions this by way of a scenario. He says: Imagine that I am free to choose between A and not-A and choose A. There is a meddler who could have coerced me into choosing A had I chosen anything else, but he did not actually do so. Thus, it appears that we can praise or blame people who cannot really choose other than they do.

There is one new thing in Frankfurt's scenario. According to believers, God meets two criteria for a meddler: he reviews our decisions and he could interfere with them. The only difference is that God's non-interference is not conditional on my having chosen what I did: God refrains from interfering no matter what we choose to do. This is because he wants us to have free will. The only thing omitted is a god who would have interfered had any of our choices been other than they were.

Frankfurt's scenario does extract a concession but it is a trivial one: it makes a difference just how our alternatives become limited to one. To undermine praise or blame, it must be by way of an active meddler. The culprit that dictates my choices could be either an interventionist God or a facet of brain physiology. When either one substitutes for the present self as decision-maker, it abolishes praise and blame. On the other hand, a passive power that limits the present self to one alternative leaves praise and blame intact. What is important is not the brute fact that I have been left with only one alternative: it is how that occurs. It must not be by way of interference with the autonomy of the present self.

Another mode of leaving me with only one alternative is when the present self makes a choice and acts on it. The past cannot be altered. The only way to keep alternatives open would be a Hamlet-like indecision that lasted forever. Choice is a way of restricting alternatives that does no harm because it does not involve active interference.

The universe in which I actually live has no passive meddler. If determinism is true, it is totally dominated by active meddlers. Therefore, its way of limiting my alternatives to one really does render praise and blame absurd. For his concession, Frankfurt has paid a huge price: he has granted that if active meddlers are present, praise and blame are indeed abolished.

I argue that if everything, including our apparently free choices, is subject to scientific explanation, praise and blame disappear. Frankfurt's scenario does not challenge this. An eternally passive cause would be unobservable, and therefore could not play a role in scientific explanation. I believe I know the identity of the meddler: it is the organ called the appendix. I believe the appendix has the power to meddle with all our choices were they other than they are but has never done so. Thus medical science has made the error of declaring it a functionless organ. The first day the appendix acts it will be observable. However, the day it acts as an active cause, it will also abolish free choice.

It may be said that science has never observed brain physiology meddling with our choices, but physiologists are convinced that someday they will able to show that states of the brain act as prior causes, dictating outcomes. An "agent" that might have intervened simultaneously is not the same as a prior cause. Science has no room for a

SCIENCE AND FREE WILL

The point can be put as a syllogism:

  All choices with active causes are not free.

  A choice science explains has active causes.

  Therefore, a choice science explains is not free.

Or:

  All choices made without determination are free.

  A choice subject to veto is made without determination.

  Therefore, a choice subject to veto is free.

never-active meddler. Compatibilism is defensible only when a scientific perspective is abandoned. That bodes ill for its viability (see box above).

A paper by Peter Strawson (1919–2006), "Freedom and Resentment", is much cited. Strawson points to the reactions we have when someone injures us—for example, resentment, expecting them to be sorry, falling out of love with them. Should we give these up if we believe in determinism?

His first answer is that we cannot because these reactions are too deeply ingrained. He anticipates this will spark the reply: Even if that is so, it evades the question of whether, if rational, we will see that they are no basis for a judgment of moral condemnation. His rebuttal is that whether we give up our reactions is a practical question, in no way dependent on theoretical questions such as the truth of determinism. We should assess whether we would benefit from giving up our reactions and look at how this would impoverish our lives.

The idea that theory should not affect practise is so odd that other disciples of Wittgenstein turned the argument into a language game. For them, to say a person is free is not a description of what people are. It is just a way of signaling that we are prepared to react to them with indignation and so forth.

Whether the argument is psychological or verbal, it is bankrupt. It makes it legitimate to assess my belief in God in terms of whether I benefit from it. Or it assumes that when I say, "I believe in God" this merely signals that I am prepared to be worshipful. We must hope that we never encounter a people who deeply value blaming (or worshiping) clocks.

Wittgenstein (1889–1951) asked what value being free has in the sense of one being able to do other than one does. I refuse a bribe. What would it mean to say that I could have accepted it? I could do so only if I were morally corrupt, and who would want to be told that?

In rebuttal, this is a case in which I did something right. The sterility of being able to do other than one does is less clear if I have done something wrong. Let us say I took the bribe. Would no one welcome the possibility that he or she could have behaved like a better person? Would no one prefer to believe that the present self had some influence here, rather than believing that this kind of behavior was beyond its control?

Did Wittgenstein never regret anything he did? He was so arrogant that he turned every session of the Cambridge Philosophical Society into a monologue. He ruined the lives of students by advising them to abandon philosophy, which he adjudged futile, and become laborers. He wished to add his personal bit of killing to the First World War, a war that had little honor on any side. If he did not welcome the possibility of alternatives to his behavior, many others will wish his "character" had been free to choose otherwise, and that he had actually done so.

These philosophers of language do not analyze it very well. The following is a typical example of moral discourse. Imagine I steal something from a friend. He confronts me and says, "I am disappointed in you." Under the presumption of my freedom, the meaning is quite straightforward: "You and I both know you could have done the right thing." It is a clear indictment that I am morally blameworthy.

From Wittgenstein's perspective, the assertion becomes convoluted. My friend should say, "I am disappointed with myself." After all, if my friend had properly assessed my character, the theft was predictable. He is not disappointed in the real me at all. He is disappointed in an illusory me who never existed. He was surprised at my decision only because of self-deception about my character. Whose fault is that? Certainly not mine.

His assertion really means "I now see you as you really are". He can soften his words by adding "but one theft does not make you Jack the Ripper". However, this does not obviate the fact that he has purged his assertion of moral blame. He now sees he can expect worse acts on my part than he suspected, but he has been robbed of the ability to say that I made the wrong choice. There was no free choice.

Steven Pinker (born 1954) asks the question: Would anyone want people to be free to do anything whatsoever? If that were so, neither reward nor punishment would affect human behavior. Nothing would be an effective incentive or disincentive.

Pinker's question ignores how the present self operates in making free choices. After being consistently fired for my Social Democratic politics in the United States in the early 1960s, I was free to do anything within my power: stay and keep getting fired; abandon academia; commit suicide; go overseas. I eventually eliminated all but the last option because of certain considerations, mainly that I wanted to live, had a family to support, and had an intellectual curiosity about certain things. As for going overseas, I earn my living through talking so that left English-speaking nations—Britain, Ireland, South Africa, Canada, Australia and New Zealand. Alarmed by the possibility of nuclear war, it seemed to make sense to choose a remote area such as New Zealand or Australia. As a Social Democrat, I had a much more positive image of New Zealand than of Australia because of the latter's white Australia policy.

My description of this process in no way implies that my formed character left the present self only an executive role. There were temptations to resist in every direction; they included self-pity, love of the familiar, and the higher salaries available in Australia. I had to choose between alternatives that both my character and the real world left open. However, even under the presumption of freedom, the outside world influenced my choice. Universities kept firing me, certain nations had adopted English as their dominant language, some had done things to make themselves less likely targets for nuclear destruction than others, and Australia had compromised its social democracy more than New Zealand had. What others do structures the choice of free present selves, even if their actions pose alternatives to be weighed, rather than dictating what people do.

Compatibilists always tell me there is someone new I must read before I make up my mind. After being told to read Strawson, I gave up. I believe that either free will or determinism is true, that free will makes moral praise and blame appropriate, and that determinism makes them inappropriate.

WHAT IF BOTH POSSIBILITIES ARE OPEN?

I realized that if everything depended on whether free will or determinism were true, the arguments for and against free will had to be taken seriously. But then I got a surprise: none of the arguments were decisive. Worse, we would be in doubt as to what was true for the foreseeable future. Therefore, we needed a substitute for the truth that we could use to live by.

My first task is to defend the assertion that none of the arguments for or against free will are decisive. As a sample, here are five arguments—three for and two against.

First argument against free will: Everything that occurs in consciousness is linked to a physiological state of affairs in the brain. The two interact, and if causal discontinuity occurred on the level of consciousness there would be a corresponding causal discontinuity in brain physiology, which is absurd.

Answer: What is absurd about it? This argument assumes that the brain operates like a clock, and if a capricious consciousness intervenes the machinery will be disrupted. But if the brain is a clock that operates best when left alone, *any* influence from consciousness may be disruptive. If a "decision" has been arrived at after being determined, why would it be less disruptive than a decision arrived at freely?

If I take drugs, the drugs will affect my brain physiology, no matter whether the act was free or determined.

The whole picture of brain and consciousness as two alien entities is a throwback to René Descartes (1596–1650). The brain influences consciousness and consciousness influences the brain. Consciousness has free choice only when an underlying brain state allows for that, and after consciousness makes a free choice the brain reacts to it.

Second argument against free will: Brain physiology has already falsified free will. The British physiologist W. Grey Walter (1910–1977) reported an experiment he conducted in the early 1960s (although he never published his data, leading to speculation about why he was reluctant to do so). Electrodes were inserted in the motor areas of the brains of epilepsy patients. Walter ran wires from the electrodes to a slide carousel. Whenever a patient decided to move to the next slide, electrical activity in the brain beat them to it and changed the slides. The patients were astonished. They felt that just as they were about to push the button but had not yet quite decided to do so, their brains had made the decision for them.

Answer: The fact that the electrical impulse from the brain changed the slide is sheer showmanship. It could have just as easily lit up a light bulb, or simply been detected by a brain scan.

The significance of the fact that there is extra electrical activity in the brain just prior to the fruition of a decision is no more significant than if there were reduced electrical activity. All we know is that something distinctive happens in the brain just before a decision is consummated. The electrical activity in question may even have shifted backward from its original position in the process. When the eye blinks in response to a stimulus, it can be trained to blink at a perception that always precedes the stimulus and would not in itself cause a blink.

First argument for free will: This comes from Jesuit psychologists, who must square their science with the praise and blame allocated in the confessional. Once consciousness evolved, they say, evolution would have tended to produce a creature whose present self had free choice because of the obvious survival advantage of this. In situations where there was time for reflection, a creature that could choose from a huge range of responses to challenges from the environment would be more successful than one limited to only a certain set of responses.

Answer: A deterministic system can also generate a huge range of responses. The immune system creates an enormous number of antibodies that are unspecialized: they are not limited to fighting one kind of disease but act as tiny "Darwin machines". When the body is attacked, the immune system selects the antibodies most suited to fighting the infection. Some become specialized and give immunity to a particular kind of illness, but plenty more are left general to fight the next unknown enemy. There is no reason to think the possible responses of a determined consciousness are any less vast.

In addition, the Jesuit argument assumes that evolution is actually capable of producing free choice, which may or may not be true. True free choice may be optimal, but perhaps the best that matter can do is to produce the illusion plus a huge range of determined responses. The illusion of free choice must have survival value or it would not have evolved, but it may still be an illusion.

Second argument for free will: A rational mind escapes causality. When you use a syllogism to arrive at a decision, conscious reason is in control.

Answer: Using a rational tool does not imply a present self that is capable of choosing to do this rather than that. A combination of genes and environment may produce a consciousness capable of reason, and one so committed to reason that rejection of the irrational is automatic. In this case the laws of reason dictate the choice. True freedom means that the present self commands, in the sense that it decides to forgo the irrational, however tempting it may be, in favor of a commitment to reason; without that "effort of the will" the irrational would prevail.

Third argument for free will: If I could raise my right hand and then go back to that same moment and raise my left hand, the truth of free will would have been proved. We cannot do this but we can come very close—that is, we can raise the left hand after an infinitesimal amount of time has passed. Is it really plausible that something altered in the intervening one-thousandth of a second to determine a different outcome? The determinist is driven to positing a mysterious factor X that must have been added to the causal mix. What was it—the motions of Jupiter, or perhaps some slight alteration in the blood supply to the prefrontal lobes?

We could reject the outcome of any experiment on such grounds. After experimenting with mixing hydrogen and oxygen, we keep getting water. That is, we get the same outcome. It is possible that a factor X is present that dictates the uniformity of the results and therefore we should not trust them. Who would take this seriously? Why do we reject *differential* outcomes under the same experimental conditions in which we accept *same* outcomes, save that we have an irrational bias in favor of determinism?

Answer: Assuming the mind is a determined system that produces set outcomes, there is no reason to think that it is less able than a series of free choices to dictate the pattern of a series of decisions. It can dictate right–left–left at ten p.m. and right–right–left at just after ten. The alteration in brain physiology between the two times would hardly have to be great. The argument ridicules the possibility of a factor X and then proposes one. The thing that has changed in a fraction of a second is supposed to be an altered mental state: deciding to do A rather than B.

If no argument or evidence offered so far decides the status of free will, what *may* count as evidence? How can science decide the question of whether its own sway is unlimited, or is circumscribed by the existence of present selves creating uncaused outcomes? I suspect it can do this only if brain physiology comes to the rescue.

Someone sits in a room with all the readings imaginable from another person's brain. He or she notes the characteristic reading that signals the beginning of a decision-making process and follows the incoming data though to the reading that a decision has been made. Analysis of the data (which may take weeks) allows them to correctly describe the decision. For example, the person was deciding between going to see a film or visit a sick friend and decided to visit the friend; the person decided to hold his tongue at a meeting of the Cambridge Philosophical Society so others would have a chance to speak.

It would be wrong to demand a one hundred percent success rate. What would be interesting would be if most decisions could be predicted but that there seemed to be a boundary around a class of decisions difficult to breech, where the present self was really torn and had to make an effort of the will. If a "perfected" physiology found it could post-predict most decisions outside that boundary and few within, free will would be more probable. If that boundary proved irrelevant to post-prediction and was consistently invaded, determinism would be the more probable.

There is a problem with the above scenario: it has our physiologist receiving brain readings only. The mind is more than the brain. It is a functional system with both brain and consciousness as components that influence one another. Therefore, in order to moderate a mind through the process of decision-making, we would have to have consciousness readings as well as brain readings. The subject in question could hardly report to the physiologist as to what they were thinking. To do so up to and including the decision would give the game away. Moreover, can anyone accurately report all that is going on in their consciousness? And even if they could this kind of reportage would be a strange addition to normal consciousness.

If knowing the present state of another's mind is impossible, and if that is a prerequisite for predicting apparently free choices, we will never know the truth. We will know that either free will or determinism must be true but never have a rational guide as to which is true. Even if this is too pessimistic, brain physiology will not settle the question in the foreseeable future.

How are we to find our way about in this strange world in which what we need to know to live our lives is unknown? You may ask whether we really need to know. I will argue that we do because we must all decide whether or not to play the blaming game.

Here is our predicament. Imagine three identical doors. Behind one is a universe where a certain free decision has been made—namely, to pick up hitchhikers—so it is in state A. Behind the next is a universe where another free decision has been made— that it is too dangerous to pick up hitchhikers—so it is in state B. Behind the third is a universe in which no decision is free, in the sense of creating causal discontinuity, and it is in state X—that is, in whatever state causality has dictated. That could be either A or B but of course it cannot be both.

We go through one of the doors but see nothing to tell us which universe we live in, at least for the present. Therefore, on one level we should suspend judgment as to whether we are in a free or determined universe.

On another level, however, we cannot suspend judgment: when we interact with other people we must decide whether or not to play the blaming game. We must decide between two ways of judging people. We can approve or condemn them only as causally responsible for good or bad conduct: "You have done something wicked and it was you who did it and not someone else." (Other clocks chimed on time.) Or we can praise and blame them for the outcome of free choices as well: "You know very well you could have chosen differently." (You could have chimed on time.)

We must decide whether the moral indignation we feel when someone deceives us about the time is any more appropriate than when a clock deceives us about the time, and the blaming game makes sense only if our universe includes free choices.

In sum, we know that the universe is indeed free or indeed determined, we have no notion of which, and yet we must still choose, despite the absence of rational guidance. We must simply decide whether to play the blaming game.

This choice is unique. It is the sole existentialist dilemma: we must choose "policies" based on one or the other of two pictures of reality without any rational guidance. Other decisions about belief are made on the basis of reason (whether something other than ourselves can tell us what is good), evidence (whether Earth is round), or logic (whether triangles have 180 degrees).

Since reason cannot help, what about survival value? Believing in free will rather than determinism may be more advantageous, so would we not do well to follow this course? However, what is advantageous is not the belief, it is behavioral patterns that come readily to hand from our evolutionary past. For example, assume that moral indignation has survival value for me because its display convinces potential aggressors that I will take a terrible revenge. If that is so I will not try to suppress it: the belief in free will that underpins it may be true.

However, the survival value of moral indignation should not convince me to believe in free will. Many species, such as fish that puff themselves up to look larger and fiercer than they are, have deceptive displays that help them survive. I doubt someone who is agnostic about free will has any difficulty showing emotion when attacked, although I suspect that there are deterrents more effective than displays— namely, a history of terrible retaliation against aggressors that will give any new enemy pause.

If I am correct, every one of us faces this choice: we must assume either free will or determinism, and must do so without any proper guide from the intellect.

This doesn't mean the decision is subject to no regulation. Since it has consequences for yourself, others and society, it is a moral decision. I believe it would be dehumanizing to treat my intimates as actors beyond moral praise or blame. If my partner did something thoughtless and hurtful to me, I would judge her to have freely chosen to do something she could have refrained from doing and I would react accordingly. For all I know, she may indeed be culpable. My censure does not rule out license for human frailty—we all choose wrongly sometimes—so I would be ready to accept an apology.

I frankly feel that to drain from personal relationships the dimension of praise for choosing well and blame for choosing badly would be a charade on my part. I could tell myself I was doing it—but only because I did not really feel that I was doing it.

However, I am not so arrogant as to assume that those who adopt a different policy are acting in bad faith. It would be just as rational for a couple to regard each other's behavior as determined and they might find that amenable. Both would know that a display of moral indignation should be interpreted as strong intolerance of the substance of their partner's behavior. Both would know that an apology should be interpreted as a claim that the behavior in question was atypical and unlikely to occur again.

When we choose to play the blaming game, morality imposes its own logic—namely, that we must apply moral rules with consistency. My partner has every right to demand that if I play the blaming game with her I do not exempt myself and tell her all my choices were dictated by factors outside the control of my present self.

But logical consistency does not forbid a different policy where the morally relevant circumstances differ. Our penal system in certain areas is evolving towards the elimination of fault in the moral sense. Trying to establish whether husband or wife was at fault simply embitters divorce proceedings, and is best set aside in favor of an equitable division of property and the welfare of children.

From a humane point of view, we may wish to further this trend and seek only protection of the public and reformation of the criminal, with punishment for punishment's sake, punishment that matches the wickedness of the choice, set aside. For all we know, the behavior of those who transgress really may be determined and moral blame inappropriate.

On the other hand, we must be mindful of the feelings of others. When the evil consequences of behavior are very great or the personal damage horrific, it may be too much to ask the injured party to accept a legal system drained of righting the moral balance sheet. Should Jews treat Hitler's behavior as determined? Should I treat someone who raped and killed my daughter as having no choice in how he acted? It could be that the horrific acts in question were the products of free choice.

The analysis developed here allows maximum flexibility. Since the truth of free will is unknown, we are free to include in our legal code whatever mix has the most humane consequences. Once again we cannot change the mix from day to day: people have a right to anticipate what penalties the law will apply.

Although I play the blaming game with my intimates, I may not play it in other social roles. I may be a psychiatrist of the school that believes personalization of the tie between alienist and patient is counterproductive. Therefore, I treat my patients as creatures whose behavior is beyond their control, and weigh whatever I say purely as a means of enhancing their chances of recovery. If I have a son who is showing psychopathic tendencies, I may decide it is better to adopt the policy of the alienist rather than the one I apply to most of my personal relationships.

When I play the role of social scientist, I will certainly assume that determinism is true. After all, it may really be true. Any other assumption could set premature limits on scientific explanation. Let reality set those limits: it is not my job to anticipate them. As for judges, their role depends on whether the legal system deems moral culpability relevant to sentencing in the case at hand.

I am free to praise or blame others in every case in which there is doubt about whether their behavior was determined. It goes without saying that there are multitudes of cases that do not qualify, ranging from the decisions of infants to the decisions of youths where circumstances make certain outcomes, such as being socialized into a gang, virtually automatic. Even in the present state of our knowledge, we know that these "choices" are not free.

We can decide whether or not to play the blaming game in a way that accords with our moral principles, but the choice is open only because of our ignorance. It is infuriating not to know the truth about free will. Living the examined life is not always a piece of cake.

ARE SPONTANEOUS OCCURRENCES POSSIBLE?

When I was in high school it never occurred to me that science might lack a rational foundation. After reading David Hume at university, I could think about little else. Hume showed that science rested on an act of faith—namely, that spontaneous occurrences are less likely than lawful occurrences.

By this time I was an atheist so I found it intensely annoying to encounter religious people who defended their faith by arguing that science was just as much a matter of faith as religion was. It was not that I was tempted to accept a religion. Telling me I must have faith in something does not tell me that I must have faith in anything in particular, such as Catholicism, or Calvinism, or Buddhism, or Islam. But the claim that those I regarded as irrational were on a par with me struck me as false. It became important to be able to pinpoint why.

Philosophy rests on a mix of metaphysics (what things exist) and epistemology (what assertions are true). Neither can do without the other. On the one hand, unless you believe God exists it hardly makes sense to pursue theology, the science of what it is possible say about God. On the other, you should not believe God exists until you establish the truth of the proposition that asserts his existence.

A basic epistemological problem must be addressed before you can formulate any theory of being: the problem of induction. The core of this problem is that we have no reason to reject the occurrence of utterly spontaneous events. Since all theories of being posit the persistence of existing entities and causal relationships, none can survive the occurrence of spontaneous events. In addition, no metaphysics has a solution to this that does not beg the question. Indeed, I can see no solution from any quarter.

Therefore, we have a choice: either set the problem aside or espouse no metaphysics at all. If we set the problem aside, we can address the question of what theory of being is most defensible from scratch. When that is done, science, I believe, emerges as the only valid method of establishing what exists. I will try to convince you of that later.

Induction is different from deduction. Take a syllogism such as: "All men are mortal; Socrates is a man; therefore, Socrates is mortal." Once the two premises are established, deductive logic tells us that Socrates is mortal.

But what of the main premise, that all men are mortal? That is a product of induction, by which we mean it is a generalization from experience. Up to now, every person who has lived has eventually died. The assumption that we can make reliable generalizations from experience underlies not only everyday reasoning but also the whole of science. We have collected a huge body of observations that show, for example, that Newton's law of gravity is roughly correct.

Every scientific law assumes that human experience is a good voucher for generalizations. Moreover, the very point of science is to allow us to predict future experiences on the basis of past ones. We rely on Newton if we want to anticipate where we will see the planet Jupiter tomorrow.

The Scottish philosopher David Hume (1711–1776) posed the problem of induction when he examined the foundations of science. He showed that past experience engenders no rational expectations about future experience.

His analysis has four steps:

One: No one can get into a time machine and observe the future.

Two: Deductive logic says nothing about the occurrence of experiences, even in the present. If you tell me there is a yellow balloon floating in the hall outside my study, I have to go and look. I cannot deduce from logic whether it is there. If logic cannot tell me about present experiences, it can hardly tell me about what future experiences will be like.

Three: We never perceive necessary connections between events. If we could, things would be different. If event A were necessarily connected to event B, then the occurrence of A in the present would justify the anticipation of B in the future.

Four: These alternatives exhaust all possible ways of arguing that future events will reflect past events. Since all three fail, we have no rational warrant that the future will resemble the past. All of the observations that the sun has risen each day tell us nothing about whether it will rise tomorrow (see box on page 141).

You may think Hume was too hasty in asserting that we never see necessary connections between events. However, he is correct. Imagine we have with us some people from a pre-industrial society. We pass a magnet over some iron filings on a table so the filings leap up to the magnet. If our visitors could see a necessary connection between the magnet and the movement of the iron filings, one experience would be enough. They would have seen a necessary connection and that would be that.

In reality, though, these people will be amazed and suspect trickery. They may look under the table to see if some device is flipping the iron filings upward, and inspect the magnet to see if there is some kind of glue on it to hold the filings when they get there.

We would have to repeat the sequence of events several times in order to convince them. When we did so they might finally conclude that there was a necessary connection, but they would not have seen one. And if Hume is correct, they would not have the slightest reason to assume the magnet would work next time we used it. Nothing they saw would justify the assumption that the future would tend to resemble the past.

Some are confused about what Hume showed. They claim he proved only that certainty about the future is impossible, which leaves probability intact. They assume that if there is a lot of evidence that B has followed A over and over again, it is at least likely that it will do so in the future.

HUME AND WOMEN
Hume found the problem of induction deeply depressing. "I dine, I play a game of backgammon, and am merry with friends; and when after three or four hours' amusement I return to these gloomy speculations, they appear so cold and strained and ridiculous that I cannot find it in my heart to enter into them any further," he wrote in A Treatise of Human Nature. He also said he enjoyed the company of attractive women. His reputation as the pre-eminent philosopher of the English-speaking world rests on his gloomy speculations: philosophers think them more important than his taste in women.

This assumption is logically incoherent. It treats past experience as if it were a sample from a bag containing both past and future experiences. If Hume is correct, we have no reason to think we can put past and future experiences in the same bag. A sample composed of women does not allow you to generalize about men except insofar as the sexes are alike. We have no reason to believe that past and future share any commonality at all. You cannot make probable statements about butterflies based on a sample of lizards.

It may be said that science does not actually assume necessary connections, but simply tests its generalizations against experience to see whether we can falsify them. As long as the generalizations and theories survive the tests they are worthy of qualified belief until a falsifying experience comes along. Hume's point is that they are not worthy of qualified belief or any kind of belief insofar as the future is concerned. Suppose that up to now we have never found a black swan. However, the next time we look at a swan it may well not only turn black before our eyes but also become a black squirrel. If there is no continuity between the future and the past, why not? Past failures to falsify tell us nothing about the future.

Finally, there are those who think they can conjure the problem away with words. They assert that basing expectations about the future on past experience is what we mean by being rational. Irrational people base their expectations on their hopes and dreams, or on astrology, or on some other theory for which there is no evidence. This is sheer question-begging. If Hume is correct, it is no more rational to base expectations on the past than on anything else, because no rational case whatsoever can be made for doing so. It is hard to be more bankrupt rationally than that.

This rebuttal may seem plausible: In the past, predicting the future on the basis of the past has always worked. Here we have a perfect example of circular reasoning. The point at issue is whether the past is indicative of the future, so we can hardly cite the fact that, in the past, the past has always been indicative of the future.

Presumably no one will cite the concept of "laws of nature" as an answer to Hume. If such laws exist and are permanent they would certainly guarantee continuity from past to future, but Hume's whole argument challenges our right to posit laws of nature. What is a law of nature except something that gives the events of the physical universe a necessary connection—or at least a probable connection—with one another? The whole concept is a casualty of the problem of induction.

All these evasions ignore the core problem, which is that we have no reason to assume that utterly spontaneous events will not occur. Indeed, we have no reason to even assume that the identity of objects will continue into the future. Therefore, I am going to rename the problem the "problem of spontaneity". This has the advantage of freeing it from the context in which it arose.

Previously it has seemed a problem peculiar to science. I will now show it is only a special case of a larger problem of spontaneous occurrences, that this larger problem affects all theories of being, and that advocates of non-scientific metaphysics are no more able to solve it than are philosophers of science. Rather than address all non-scientific metaphysics, I will discuss Aquinas, Plato and Kant, and presume that if they are at risk lesser thinkers are too.

If everything in the physical universe is capable of spontaneous occurrence, God's role as creator is superfluous. As Thomas Aquinas (1225–1274) said, God's omnipotence requires only that he has the power to create a universe, and not that he has actually exercised that power. To be an omnipotent carpenter you have to be capable of making a perfect chair and as many as you wish, but you need not have made any as of this moment. This implies that the universe that exists may have originated spontaneously and a God-created universe still awaits us.

There is a deeper problem that has to do with Aquinas's analysis of the nature of God. God, he says, is a necessary being in the sense that he requires no cause of his existence, unlike all lesser beings, who depend on God for their existence. The problem of spontaneity suggests that all things can pop into existence quite independently of God.

The problem of spontaneity is fatal to Aquinas's concept of God and his concept of the relation of God to the physical universe. The problem takes on a special character according to what metaphysics it undermines, and I shall call this special case "the problem of creation".

As discussed earlier, Plato believed the fact we can classify things that exist in the physical universe posed a logical problem that we could solve only by positing an abstract concept for each class of objects. There had to be a general concept of chair broad enough to include all particular chairs, not only those that existed but all those we could imagine.

He also proposed abstract concepts of morality (justice), relationships (equality), mathematical notions (triangle) and, by extension, time and space. Each of these abstractions bore a one-to-one relationship with existing entities he called forms—that is, a form of chair, a form of triangle, and so forth. They were no less abstract than the general ideas, but unlike them existed independently in a world of forms.

For Plato, the forms had a causal role in shaping the physical universe. Prime matter tends to lapse into formless chaos, a sort of unstructured mess like cookie dough before molds stamp out any cookies. The ordered physical universe we see around us gets its structure because of the influence of the forms. The forms "tug" the stuff of the universe into an orderly process of development, rather like magnets bringing order to a jumble of iron filings by aligning them along the lines of force.

However, since the forms are changeless they cannot create a tendency towards one change rather than another— that is, they can do nothing to determine that this will occur as distinct from that. Since prime matter is without structure, it cannot dictate anything except a tendency toward chaos. Therefore, even if Plato's theory of the physical universe were valid it would not solve the problem of induction.

The problem of spontaneity also poses difficulties for Plato's particular metaphysics. His metaphysics is like Christianity in the sense that both assume a special relationship between something else and the physical universe in which we live. Just as the possibility of spontaneous occurrences sidelines God's creative role as author of the universe, it sidelines the creative role of the world of forms. If events that take place in the physical universe require no connection to anything, it is superfluous to propose that the forms have a role in shaping those events.

It may be said that both Christian and Platonic meta-physics posit an implicit rule that such an event cannot happen. Well then, a scientifically based theory of being can pass its own rule. All that is required is the assumption that events can occur in the physical universe only if they obey unalterable natural laws. Solving the problem of spontaneous occurrences by legislating that they cannot happen is a "solution" to the problem available to anyone.

The relationship that the forms are supposed to have with the physical universe is fundamental to Plato's metaphysics. When this tie is severed, the result is destructive, no less so than when we severed the relationship between God and the physical universe that Aquinas posited. I will label the special character that the problem of spontaneity manifests in the context of Platonic metaphysics "the problem of formal impotence".

Kant believed that we see whatever exists through certain spectacles that are embedded in our nature: if we see the world through pink spectacles, everything will look pink. In fact, we see it through the spectacles of space, time and causality. If these spectacles were permanent, we would always see the universe with a particular spatial orientation and its events linked with one another (the causality spectacles). However, if all things can alter spontaneously, the structure of our spectacles could alter. Therefore, we can have no expectation that we will see things oriented in space-time at all, or that we will perceive one event connected to another: once again we may have formless chaos. I will call this "the problem of unstable spectacles".

Let us talk about the duration of events. If a swan "suddenly" becomes a squirrel, is the event instantaneous or does it have duration? We would be able to perceive only that it had occurred from one moment to the next. But that would also be true of the *failure* of a spontaneous event to occur—that is, we could perceive only that a swan remained a swan from one moment to the next. I am unconcerned about what answer is given to this question, as long as the same answer is given for both stability and change: either both take time or both do not.

This digression is necessary to show that Kantian solutions to the problem of spontaneity beg the question. For example, since the Kantian spectacles "create" the time we experience, it may be said that they themselves exist outside time. After all, we do not see our spectacles through our spectacles, so how could our spectacles have a temporal orientation? And if they do not have a temporal orientation, how could they change?

The argument assumes that by putting our spectacles outside time we have given stability an advantage over change, but that has been shown to be false. If change is not possible outside time, neither is stability. And if change is an option, we need merely rephrase Hume's argument using non-temporal language. Given a population of non-temporal entities, none of them can be assumed to have a fixed identity because neither logic nor observation nor anything else could vouch for that.

Kant has only two options. Either he asserts that things can remain stable or alter outside time, with no rational guide as to which is true. Or he asserts that things cannot be said to remain stable or alter outside of time. Neither alternative gives an answer to Hume. You may object that stability "outside time" just seems more conceivable than instability, but that is true only because we tend to cheat. We imagine an entity unaltered while a few seconds of subjective time go by. That is, without conscious effort we slip on our time spectacles.

It is worth noting that our only reason for supposing our space spectacles are three-dimensional rather than twenty-three-dimensional is based on our past experience. And if what I experienced tomorrow did alter, I would be forced to posit a different kind of "timeless" space spectacles. The notion that the character of our spectacles is fixed, in a way that the experiences we see through them are not, is fundamental to Kant's metaphysics. Thus the problem of spontaneity is devastating.

Worse still, Kant's theory of being has a feature that makes it peculiarly vulnerable. Kant suggests there is a reality beyond our spectacles that is unknowable. If it is unknowable, we cannot know that it is immune to radical change. And if it makes a contribution to what we experience, then what we experience is subject to radical change.

Take a 3-D movie. When you look at the screen without the spectacles provided by the cinema, you see it as two-dimensional and fuzzy, but with the spectacles on you see it clearly, and as three-dimensional. Now imagine that you could never take off the spectacles. All you could say about the screen would be to state the brute fact that it existed. You would have no idea whether or not it could alter. Now imagine that the screen does alter: it suddenly becomes the kind that exists for ordinary movies. What we see through our spectacles suddenly alters too. There is no longer any 3-D effect. The screen-in-itself is not purely passive but makes an active contribution to what we experience.

Kant seems to have left open the possibility that events on the level of things-in-themselves affect the world we experience. He tells us that since we see human behavior through cause-and-effect spectacles, we perceive every human act as caused. But he adds that for all we know the will-in-itself, the will that exists beyond what we see through the spectacles, is free. Now, unless this will affects our behavior, its choices will be of no consequence. And if what the will does on the level of things-in-themselves has consequences in the physical universe, then other events on that level must have consequences in the physical universe. This means that if the objects beyond our experience alter radically, our perception of events in the physical universe will alter radically.

The will-in-itself is beyond time. When it makes its free choices, are these events instantaneous or do they have duration? If they have duration, the whole concept of a will beyond time implodes. If they are instantaneous, why shouldn't changes in the structure of our spectacles also be instantaneous?

Three things may be concluded from all this. First, all metaphysics suffers from the problem of spontaneity. This generates special problems specific to particular kinds of metaphysics: the problem of induction within any theory of being based on science; the problem of creation within theology; the problem of formal impotence within Platonic metaphysics; and the problem of unstable spectacles (and indeed unstable things-in-themselves) within Kantian metaphysics.

Second, since the problem of spontaneity manifests itself with equal destructive force within all kinds of metaphysics, scientifically based metaphysics has no monopoly on it and is at no special disadvantage.

Third, if the unsolved problem of spontaneity renders all metaphysics non-viable, you must either set aside that problem or have no metaphysics. This is not a matter of inventing a new evasion by playing with words. The penalty of "no metaphysics" is not an excuse to ignore the fact that we have no reason to anticipate anything (other than utterly spontaneous events). It merely supplies a motive.

To answer the question posed at the beginning of this chapter, yes, science does rest on faith. However, the only act of faith required is to believe it is possible to make any sense out of the universe at all. We must ignore the possibility that the universe admits occurrences that are utterly arbitrary. And religious belief (in a god who created our universe) must ignore that possibility too. If you approached the future as capable of presenting you with any possible event, however horrific, you would be immobilized by terror. If you read the past as something that could have been the product of utterly arbitrary events, you lose God the creator. Some of you may believe you have a solution to the problem, that there is an infallible book that tells you God exists and created the universe. I will deal with this in the chapter "Does God exist?".

I propose that everyone, from Christians to Platonists to Kantians, now sets aside the problem of spontaneity in all its manifestations. That includes, of course, allowing scientists to set aside the problem of induction.

What Exists?

DOES SENSE EXPERIENCE TELL US WHAT EXISTS?

Having set aside the problem of spontaneous occurrences, I could develop a metaphysics, a theory of being. After some reflection I embraced realism. This is what I believe. One: The objects we see in everyday life (such as tables and trees) exist independently of our experience. Two: The things we detect when we use instruments (such as microscopes, telescopes and atom smashers) exist independently of our experience. Three: The history of science has taken us closer and closer to an accurate picture of the real world. Four: God does not exist. Five: Nothing "like" God, such as the Absolute perceived by some mystics, exists.

Almost everyone who has not studied philosophy just assumes the objects they see every day exist independently of their personal experience. It is a truism, of course, that we are limited to our experience in the sense that we cannot experience anything beyond it. But the question is, how should we interpret what we experience? Some things we see—say, a tree in our backyard—are still there to be seen the next morning when we wake up. Other things—say, a dream—are around only when we experience them.

It seems reasonable to interpret our experience of the tree as seeing an object that exists independently of our experience. And yet philosophers, from Thomas Hobbes (1588–1679) to George Berkeley (1685–1753), qualified this kind of realism out of existence, or at least turned it into something very odd. Strangest of all, this development began with the rise of modern science, which eventually showed us that the universe had existed for billions of years before anyone was around to experience it. Nonetheless, the current of thought was so potent that it rejected a realist interpretation of science itself. Even today, philosophers such as Bas van Fraassen believe science is just a way of making sense out of what human beings experience.

My task is to first try and make the rejection of realism plausible by summarizing its history, and then show that the doctrines that deny realism, whether they are called idealism, neo-Kantianism, instrumentalism or postmodernism, are flawed.

Hobbes believed recent science showed that everything in the universe was composed of bodies in motion. I do not know whether Hobbes was aware of Boyle's law, which was published in 1662, but it is a good device to explain what Hobbes had in mind, particularly if we take a few liberties with its content.

Appearances suggest that ice, water and steam are different things, but in fact they are composed of the same molecules, H_2O, and differ only according to how these molecules react to temperature and pressure. If temperature is low and pressure high, you get ice—that is, the bodies are "frozen" into place as a solid. As temperature and pressure vary, the bodies get more mobile. First they flow as a liquid and then they become even more volatile as a gas.

Hobbes believed all of the apparent differences between things could be explained in this way. The only difference between an electric eel and a human being was that the atoms of one were arranged so as to allow for a functional electric charge and the atoms of the other were arranged to allow for thought. Thus, science tells us what things really are and all of our experiences are deceptive.

John Locke agreed that material objects were composed of bodies, but he pointed out that there were limits on what we could know about their *primary* properties. Even if we did not exist, the objects occupy space and are impermeable in the sense that nothing else can occupy their space. However, things like their color and the sounds they emit occur only when they interact with our sense organs. Therefore, these should be called *secondary* properties. I see my jacket as green but the bodies of which it is composed may be grey, and its greenness an effect of how I see it. When we say a sound is loud, how loud would it be if no one were around to hear it?

Berkeley pointed out that even seeing where an object is or bumping into it are experiences. How then can we know that objects have any primary properties at all? Indeed, how can we be sure they exist independently of our experience? The whole physical universe may be mere appearance.

BERKELEY AND GOD

Berkeley wavers on this point. It turns out that the reason the existence of things is not dependent on our actually observing them is that God is holding them in existence by actually observing them. As a bishop, Berkeley was eager for a proof of the existence of God. I assume that no contemporary thinker is inclined to follow him down this road.

Berkeley did not deny that things exist when we do not perceive them (see box). He was sure the slippers under his bed were there all night when he was asleep, a concession to realism. But he also believed that we must reinterpret what it means to say that the slippers exist "independently" of our experience. Since we can know nothing beyond our experiences, such an assertion must refer to hypothetical human experiences: "If I were to wake up and grope under my bed, I would feel a pair of slippers and then I could put them on."

This would also apply to the discoveries of science, such as that the universe is billions of years old. To say the universe existed before we did is to say that if a hypothetical person were present a billion years ago, he or she would see galaxies, stars and planets.

As we saw in the last chapter, Immanuel Kant introduced a refinement. He did not deny there was a real world beyond human experience—that is, the thing-in-itself or *noumenal* world—but he argued that all of its properties were unknowable. We have spectacles through which we experience things-in-themselves and can know only what we see through these spectacles—that is, the world of experience, or the *phenomenal* world.

This set the stage for an instrumental view of science, the idea that its role was to account for human experience and that notions such as atoms and electrons were merely useful theoretical concepts. It was not the job of science to tell us more and more about the real world but merely about the *experienced* world. Ironically, Hobbes had begun a line of analysis that would result in rejection of his own view of science, namely that it was a revelation about the real world. Subsequent schools of thought reinforced an instrumental view of science.

A. J. Ayer was a verificationist. It was meaningless, he claimed, to say that something existed unless you could spell out the experiences that could verify or falsify its existence. If someone says that God exists and cannot point to a sense experience as a test, they cannot be saying that he exists in a literal sense. At best, they are using a metaphor, perhaps using the word "God" to express a sense of awe about the wonders of the heavens.

Some followers of Ludwig Wittgenstein went a step further. They argued that since we cannot reason without language, philosophy should analyze the function of various forms of language. Stephen Toulmin (1922–2009) said the function of scientific language was to "alleviate surprise"—

that is, to anticipate our experiences and explain away any jarring ones, such as a stick half out of water looking bent. So now, rather than our spectacles structuring experience, science casts a web of language and theory over the universe that lends it structure.

In 1962 Thomas Kuhn (1922–1996) introduced the notion that science progresses by way of paradigm shifts. Kuhn took care to stress that each new paradigm is better than the previous one: Einstein's vision of a four-dimensional space-time continuum is better than Newton's vision of a three-dimensional universe with space and time as separate compartments. This, of course, leaves open the option of a realist view of science. The new theory is better at what? Perhaps it gives a better picture of the physical universe that lies beyond human experience.

However, Kuhn also said that paradigms are discrete: what is meaningful in one has no place in another, and one cannot be evaluated by another. That, of course, did not rule out testing them both against evidence. Unfortunately, the words he chose gave idiocy an entrée. The postmodernists leapt to the conclusion that paradigms could not be called more or less true at all. Thus we have Jacques Derrida (1930–2004) and the notion that reality is merely a text, a description of nothing, and that any text is as good as any other (see box opposite).

Serious philosophers of science did not go as far as Derrida, but you can probably now appreciate why there was a consensus that the role of science was merely

---

DERRIDA AND THE PIG

The Babylonians had a text about the disappearance of the sun at night and its reappearance in the morning: it was eaten by a pig and then regurgitated. The text that the rotation of Earth explains the sun's disappearance and reappearance is more adequate. Every time Derrida put on his spectacles, the text we call the theory of optics explained why the spectacles worked; no other text did. If you wish to catch a bus, it pays to have an up-to-date text (timetable) and not an obsolete one. That is enough about this stuff.

---

instrumental—that is, science was supposed to clarify human experience and nothing more.

This view of science leaves open a halfway house. Bas van Fraassen accepts that objects we directly perceive, such as tables and chairs, can be said to exist independently of experience, but he holds that the concepts of science such as elementary particles, unless directly observable, may not exist in that way. He accepts naïve realism but rejects scientific realism.

In the next chapter I will try to show that naïve realism naturally evolves into scientific realism, the existence of objects posited by scientific theories. Meanwhile, I need to defend naïve realism, the existence of everyday objects fully independent of human experience. By this I mean not only do the objects exist when we do not perceive them, but there is no need to posit hypothetical human observers.

My first proposition is that objects exist in space (and time) independently of human experience. I know this because the experiences of others position objects for me, even when I do not duplicate these experiences. Moreover, even when no one observes an object it goes on existing.

My wife and I go on holiday. The motel kitchen is dark and I cannot find the light switch. I ask her where it is and she says that she had to grope all around the wall and finally found it just behind the fridge, which stands two inches to the right of the doorway. There is enough light from the living room so that I can look there, and I see that it is just where she said.

The light switch is an object oriented in space independently of both of us. I cannot have my wife's experience of it—I cannot get into her skin and access her experiences. I did not even have to try to duplicate her experience as closely as possible by, say, groping all around the kitchen. All I had to do was follow her directions and then short-circuit things by looking just behind the fridge. The light switch was there to be found by me, or indeed by anyone else.

However, this leaves open the hypothesis that the existence of an object is dependent on someone experiencing it. This is falsified by the fact that objects alter even when no one is observing them. At one in the afternoon I take a nap and when I awake at four the hands of the clock have moved. Unless objects can function when they do not exist, the clock existed when no one was observing it.

What about the thing-in-itself? Well, the only thing-in-itself is the physical universe doing what it does when no one observes it: people are sleeping, clocks moving their hands, planets spinning in their orbits, cells dividing, tides ebbing and flowing. But, you may protest, don't we learn about all these things because we experience the universe? That is true, and one of the first things we learn from our experience is that objects go on existing when no one observes them.

What does the physical universe look like then? It does not look like anything because no one is observing it. If a waking person observes it, it impacts on his or her sense organs and looks like it looks to humans.

The fact it makes no sense to talk about what it looks like does not preclude us from constructing a model of how it functions. On every subway train in London there is graph of how the system functions that does not look like the London subway system at all. Science models many things that go on despite not being observed and whose appearance we cannot possibly know—for example, the early universe, the collision of galaxies, black holes and space-time.

My second proposition is that things have a wide range of consequences, and the fact sentient creatures can observe the things is only one of these consequences.

We can divide the consequences of an existing thing into those that affect observers (they see them) and those that affect unobserved entities (the wind ruffles the leaves of a tree). But unless their effects on observers are somehow privileged, when we say something exists we need not refer to its effects on observers. Therefore, we need not posit a hypothetical observer.

Remember, Berkeley granted that objects are there when we do not perceive them. The real debate is over the significance of this. He asserted that the only cash value of saying they are there is that they are potential objects of human experience. Just as his slippers were under his bed when he was asleep, ready to be seen and put on should he awaken, all that the existence of the light switch means is that it is there ready to be found.

I look out the window and see the wind is blowing the trees about. When I say that the wind exists independently of my experience, I do not merely refer to its potential for causing human experience. It has other potential effects, such the capacity to ruffle trees, and it would actualize those effects even if no sentient creatures existed. If we doubt this, my wife and I need merely take turns looking out the window. She perceives that the motion of the trees goes on when I do not see it, and I perceive that it goes on when she does not see it. We conclude that the wind has effects other than the experiences it causes.

Berkeley's contention that we must describe the existence of the wind in terms of its potential for human experience is logically linked to the contention that the wind's effects would not exist unless experienced by humans. Let's take a closer look at this.

Assume the wind was blowing an hour ago. Does this mean no more than that a hypothetical human observer would have experienced the wind—that is, if he or she were there they would have felt it on their face and seen the trees move? Why are we privileging this hypothetical event over any other? Why are we privileging what a human being would experience over what a hypothetical dog would experience, or what a hypothetical tree would "experience" when the wind ruffled its leaves, unless we believe that potential human experience is more intimately connected to the reality of the event than a multitude of other potential occurrences?

You may stipulate that it is mandatory to put every event into a sequence of events until you reach a human experience. For example, the wind blows, trees move, and an observer (actual or hypothetical) sees the trees move. However, why is the penultimate event subject to a demand that the ultimate event is not? If the cash value of saying the trees move is that someone sees them do so, the cash value of seeing the trees move is that someone saw me seeing it, and so on into an endless regress (someone saw them seeing me see it). I can give you as long a sequence as you want: when the trees were blown about their leaves fell on the yard, the soil became acidic, and the grass became sparse. Why must it end with a human experience?

My third proposition is that naïve idealism is democratized solipsism. The solipsist holds that only he has independent existence, and that the existence of all other things is dependent on his perceiving them. If he never finds someone who makes objects disappear when they depart, that is because their experiences are irrelevant. If I apply this to myself, the only relevant experiences are my own. When I look away the tree certainly does disappear. When I can no longer see others, they too disappear.

The solipsist has the same difficulty as the naïve idealist: when he is not observing things they change. If he travels abroad for a year and then returns, the tree in his backyard has grown, even though he was not there to perceive it. He must concede the tree has some sort of independent existence, therefore he must qualify his idealism by using Berkeley's hypothetical observer gambit: Chinese people exist, but only in the sense that if he were to go to China he would see them. However, once he grants that the existence of Chinese people has consequences independent of his perceptions—they are procreating, eating and so forth—why should their effects on a human observer (him) be privileged over all of the other consequences of their existence?

Note the kinship between solipsism and naïve idealism. The solipsist cites the fact that he experiences only what he experiences, says his task is to explain his experiences, and argues that nothing is real independently of his own experience. The idealist cites the fact that human beings collectively experience only what they experience, says our task is to account for our experiences, and argues that there is no reality independent of human experience.

I know of no argument for the latter that does not entail the former, yet idealism is taken seriously and solipsism is not. There is a kind of solipsist who is also a realist (see "Teasers" at the end of this book).

The only reason to endow a human experience with the status of privileged event–sequence stopper is that events are somehow not real until a human experience is part of the sequence. Thus, Berkeley's concession to naïve realism was not enough. When he conceded that things existed independently of actual human experiences, he conceded that we need not propose hypothetical human experiences as a substitute.

DO INSTRUMENTS TELL US WHAT EXISTS?

Science talks about things that we never see with our unaided senses but detect using instruments such as microscopes, telescopes and photographic plates. A few years ago I decided I could make a case for scientific realism, for the reality of things corresponding to certain concepts of science no matter whether the things are visible to the naked eye or only leave readings on our instruments.

We will now confront van Fraassen. This Dutch-born philosopher accepts that the objects human beings observe are real, but he argues for three propositions: one, science does not establish the existence of the things it talks about; two, we must limit claims that things are real to what the human species can observe; and three, we must do so unless we can provide a scenario in which the objects would be observable without instruments. For example, when we see the traces that elementary particles leave in cloud chambers, we should not assert that elementary particles are real. They will never be observable and thus we should remain agnostic: they may exist but they may not.

Dealing with the first proposition, it's true that science itself does not establish the existence of the entities it posits. You can take a purely empirical approach to science because the job of a scientist is to account for what we experience, and science itself has nothing to say about what is real. But this does not determine whether or not it is rational to believe in the existence of the entities science posits.

Much of philosophy is an attempt to clarify questions that science cannot settle. Clearly van Fraassen believes reason can determine what is real or not real because he accepts naïve realism: he considers it reasonable to believe that the things we observe are not just a set of actual and hypothetical observations.

The second proposition, that we should limit ourselves to what the human species can observe, cannot be defended. We use hunting dogs to detect odors we cannot smell. If hunting dogs are considered reliable, what of recent inventions such as the scentometer, which is far more discriminating than the olfactory organs of any animal? Miners used to take canaries down into mines because the birds were more sensitive to the presence of toxic gases than people were.

And here science is relevant. Van Fraassen never mentions the biological sciences, which spend much time accounting for how the perceptions of other creatures differ from our own. As we would expect, cavefish are blind, nocturnal animals are often color-blind, primates (like us) who eat fruit have good red (ripe) and green (unripe) discrimination, and coral reef fish see many colors.

We have a choice between saying that the coral reefs we observe are real and the ones tropical fish observe may not be, or that the same reefs exist for all species and different species merely perceive them differently. The jagged coral reefs, with which all creatures avoid contact, are located in the same places for fish and human divers. That is a good reason for choosing the second option. We know different species are sensitive to different parts of the electromagnetic spectrum. The part to which our eyes are sensitive we call light, but hawks can see infrared and detect urine trails left by their prey. Bees detect attributes of flowers that we miss.

Within the human species, individuals differ. Blind people use guide dogs as instruments to detect the existence of objects. Are these objects less real thanks to the fact that a blind person cannot directly observe them? If a car hits a blind person the effect is the same as if it hits a sighted person, despite the fact the blind person could not observe it directly.

Now for the third proposition, that we must limit ourselves to what the human species can observe unless we can provide a scenario in which the objects would be observable without instruments. I contend that events or objects that register on "natural" instruments—for example, wind on trees—are not privileged over events or objects that register on "artificial" instruments as long as solid theory explains why it is possible to observe some things and not other things. If that is so, it is irrelevant whether we can imagine direct observation of the object. What we "see" through microscopes, in cloud chambers, and by other such means counts as real.

How far would van Fraassen extend his thesis? He accepts as real the objects we see through telescopes. We cannot detect the moons of Jupiter without a telescope, but if we were close enough we could see them with the unaided eye. He does not say whether we should assume the existence of what we see through microscopes, but according to his criterion we should not. No matter how close we get to microscopic objects, we cannot observe them with the unaided eye.

Recent science has suggested a multiplicity of elementary particles that take us far beyond the atom. Why can we never hope to observe these elementary particles? Sometimes it is because any microscope has to bombard things with other things, either particles of light or electrons, for us to see them. These other things may be too large to delineate the small entities we are trying to see, or may collide with them in a way that alters what we want to see.

At other times, it is because these particles exist only under extreme conditions, such as the fantastic temperatures, density, and frequency and violence of particles' collisions that occurred in the early universe. When we recreate these conditions in gigantic machines, no unaided human observer can directly detect anything: we can only detect the influence of the particles through the traces they leave behind. Did these particles exist unambiguously in the early universe but exist only problematically when we simulate the early universe?

I believe we should trust instruments when we have a good, solid theory about how they detect the existence of things.

Telescopes are not unique. Magnets find iron deposits that we later dig up. Sonar instruments chart ocean depths that we send divers down to explore. Radar locates aeroplanes that later come into view. All these can all be explained by sound theory.

There are, of course, instruments that have proved deceptive—entrails of animals, divining rods and crystal balls, among others. None has a good theory behind it. There is a pattern: whenever a good theory has explained why an instrument detects objects, direct observation has confirmed the instrument's reliability. It makes sense to be agnostic about these instruments only to the degree that you doubt the reliability of the theory.

Sometimes a good theory links instruments. The theory of optics is the excellent theory behind both telescopes and microscopes, for example. Logic cannot force you to conclude that since direct observation has confirmed one optics-backed detector—telescopes—you should trust the other—microscopes—but not to do so would be like trusting one friend whose psychological profile makes him or her a reliable witness and distrusting another friend with exactly the same profile.

Once you have made the generalization that it is the soundness of the theory that explains why an instrument is a detector, there is no sense withholding detector status from another instrument that is based on equally sound theory. If what a cloud chamber detects is suspect, it is because the theory that explains why it detects what it claims to detect is flawed. If you leap the can observe/cannot observe gap, there is firm footing on the other side.

Sometimes a reliable instrument links the seen and the unseen. Van Fraassen does not discuss photography, but presumably he would accept ordinary photos that record the impression light makes on a plate: the objects recorded can be seen. Radio astronomy reveals additional heavenly bodies by recording the impressions that radio waves make on a plate. Once we know that a plate does not distort what is recorded there, should we doubt the objects of radio astronomy? I am not sure what van Fraassen thinks he would see if he got close enough to look at quasars and pulsars, but whatever he saw with the naked eye would omit much of what these objects look like on a plate that picks up not only light but also radio waves and X-rays.

There is a long list of things we detect only through theory-based instruments. What of Earth's magnetic field, which we detect from compass readings? What of the solar wind, the lethal streams of radiation that the sun sends our way and Earth's magnetic field shields us from, and which we detect by the northern lights and by the fact that compass readings go astray?

What may seem to be a highly abstruse theoretical construct has been shown to be simply a fact. Einstein said space-time warps in the vicinity of matter. It is as if you put a heavy ball on a flat blanket: the result is a sort of funnel shape with the ball at the bottom. In 2004, the satellite mission Gravity Probe B went into orbit 400 miles above

Figure 1: Composite image showing the GP-B spacecraft orbiting Earth and pointing at the guide star (higher line), while the spin axes of the four gyroscopes inside the spacecraft (lower line) are deflected by the warping and twisting of Earth's gravitational field. IMAGE COURTESY GRAVITY PROBE B IMAGE AND MEDIA ARCHIVE, STANFORD UNIVERSITY

Earth. It allowed scientists to see gyroscopes tilting as they registered the warping of space-time in Earth's vicinity. In addition, Earth turns on its axis, which has the effect of sending its funnel into a spin. In 2011, Gravity Probe B put a spinning gyroscope in orbit around Earth to measure the effects of that spin, actually "mapping" the space-time in Earth's vicinity (see Figure 1 above).

In addition, controversy about whether the mass of Antarctic ice was dwindling was settled using twin satellites in a joint United States/German mission, GRACE (Gravity Recovery and Climate Experiment). Taking the distance

Figure 2: Ordinary and dark matter separated; the pale "clouds" are dark matter. IMAGE COURTESY NASA

between the satellites as a measure, the precise strength of gravity at a point on Earth's surface was determined: less ice meant less pull. Is this use of gravitational theory to be trusted because it is detecting something about a visible object, ice, while gravity, when it detects something non-visible, such as space, is suspect?

In 2006 confirmation of the existence of dark matter struck another blow against those who reject scientific realism. Four-fifths of matter in the universe is dark matter. It would be heroic to say that twenty percent of what exerts gravitational pull in the universe not only registers on our instruments but also exists, while the other eighty percent registers but we must be eternally agnostic about its existence. The issue was decided when an actual "map" of dark matter was produced (see Figure 2 opposite).

Two problems were solved to get such a map. First, dark matter and ordinary matter are usually intermingled. However, when two clusters of galaxies collided, huge quantities of dark and ordinary matter were wrenched apart.

Second, as the Ice Cube Neutrino Observatory at the South Pole showed in 2011, dark matter is composed of Weekly Interacting Massive Particles (WIMPs). This means that it affects ordinary matter gravitationally but does not interact with light: it is dark in the sense that it is utterly transparent.

However, when dark matter is isolated in large quantities, its gravitational pull alters the position of background galaxies that are light sources. This affects the "vector" of the light they emit and "weak lensing observations" register this light; when these observations are combined they delineate the dark matter responsible. Dark matter has been both isolated and detected.

Van Fraassen accepted the existence of Jupiter because he could imagine what the planet would look like if he got close enough to observe it. Getting close to dark matter is not a problem. It is all around us because large numbers of WIMPs pass through Earth each second. If we wanted to observe a large concentration of dark matter, we could go fifty million light years away to the Virgo Cluster. A galaxy there called VIRGOH121 is made almost entirely of dark matter; indeed, it contains no visible stars. The problem would be that since dark matter is totally transparent, what would we imagine we would see?

The distinction between what is directly observable and what is "seen" only though instrument readings also encounters a host of problems in everyday life. I turn on the light in my kitchen. I cannot see the light but I infer its existence from "readings"—that is, the objects in the kitchen light up.

Actually, no one can see light of any kind, even when it is broken down into various colors. When you go to a dance and a strobe sends colors around the hall, you do not see blue light or red light in themselves. It is just that your readings of objects from moment to moment show them as red or blue. The same is true for a rainbow: when you look at a rainbow, you are not seeing light of various colors but rather moisture which is reflecting different wavelengths of light.

The detection of wind by observing its effect on trees is also a reading of the reactions of an instrument, albeit a very crude one. Inferring the existence of wind from tree readings is not in principle different from inferring the existence of elementary particles from cloud chamber readings.

In summary, our analysis of van Fraassen shows that once you accept naïve realism it is very difficult to reject scientific realism. We know what exists through a mix of direct observations, sound-theory-based instruments whose detections are subject to observation, and sound-theory-based instruments whose detections are not subject to observation. Scientific realism is vindicated.

HAS SCIENCE TAUGHT US WHAT EXISTS?

The history of science offers other arguments for scientific realism. While not strictly necessary to prove the point, it is desirable in the current climate of opinion to relate some of science's impressive features and achievements. By the 1950s a pot was boiling in European universities. In 1966 its contents were spilled on to America when Algerian-French philosopher Jacques Derrida gave a lecture, "Structure, Sign and Play in the Discourse of the Human Science", at Johns Hopkins University. The lecture did much harm. Students and their lecturers began to believe that science was in no way privileged over non-scientific "texts" or stories about humanity and the universe.

The history of science includes a long list of things that were at first invisible but later came to be seen. They include cells, seen in optical microscopes, molecules, seen in electron microscopes, and galaxies, seen in telescopes. Science has shown that the universe is far older than people thought possible. The history of science is full of such surprises. Its theories have conceived of objects and events that pre-scientific experience not only never suggested but also thought absurd. The history of science is most economically described as a process by which we learn more and more about an external reality.

Attempts to cast doubt on the existence of events in the distant past, long before any human beings were present, are not plausible. Even humble instruments can reveal the existence of events that occurred before we picked them up.

When people are recruited to be timers at athletics track meetings, they have to be briefed. The starter fires a gun at the starting line. The timers are 100 meters away at the finish line. If neophyte timers stare at their watches and listen for the sound of the gun, they will register faster times than those sophisticated enough to look down the track and see the gun's flash. That is because it takes about 0.3 seconds for the sound of the gun's firing to reach them, while light at that distance is almost instantaneous. Timers who wait for the sound start their watches about 0.3 seconds after the runners move, and a runner who actually takes 10 seconds to run 100 meters gets a time of 9.7 seconds.

If timers were to conduct repeated experiments, they could measure the speed of sound. A differential of 0.29 seconds between the two sorts of timers gives sound a speed of 340 meters a second. Other experiments would add refinements. They would show it was relevant that the timers were at sea level on a dry day with the temperature at 68 degrees Fahrenheit or 20 degrees Centigrade.

The gun fired before we detected it by the sound waves it caused. Once we have a reliable theory of sound, we could date the event (it happened 0.29 seconds ago).

Light has a speed and it is constant throughout the universe. All electromagnetic rays travel at the same speed as light. Whatever rays we get from the sun left it about eight minutes ago, so eruptions on its surface occur that long before we see them. We know the approximate distance to the stars, and we can therefore calculate the number of light-years between us and the stars. Light from the closest star, Proxima Centauri, takes 4.2 years to reach us and light from the closest galaxy to our own, Andromeda, 2.5 million years.

Recently, the Hubble telescope helped locate a galaxy 13 billion light-years away. I say "helped locate" because even the Hubble was not powerful enough to allow us to see light from this galaxy. It was located by a combination of the Near Infrared Camera and Multi-Object Spectrometer (NICMOS); Hubble's heat sensor; the Spitzer Space Telescope's Infrared Array Camera (IRAC); and a "natural" telescope, the Abell 1689's gravitational lens. This "lens" was, in effect, a cluster of galaxies roughly 2.2 billion light-years away that magnified the light from the more distant galaxy (which was directly behind it) by nearly ten times.

The sophistication of the instruments is a reason to accept, rather than doubt, the galaxy's existence. These readings mean we are seeing the universe as it was 13 billion years ago, only 700 million years after the Big Bang. Galaxies were just beginning to form in a burst of star birth that ended the dark ages dominated by clouds of cold hydrogen—an era that began only 400,000 years after the Big Bang.

We can even detect an event that occurred somewhat earlier, when the Big Bang released a tremendous burst of radiation. That radiation is still out there at the edge of the universe. We detect it as microwave radiation picked up by radio telescopes, and thereby witness an event that occurred 13.63 billion years ago.

We need not be agnostic about the actual occurrence of these events, except in the sense that as we get better and better theories of the early universe we may date and understand the events better. They occurred long before any sentient creature was around to observe them. When we say they occurred, the last thing we have in mind is that they provided potential experiences for a hypothetical human observer. We mean a huge burst of radiation occurred.

There is an unanswerable objection to the claim that the universe has a long history. About 1650, in an effort to defend scripture, Bishop James Ussher asserted that God created the universe in 4004 BC. However, rather like an unscrupulous antique dealer, he made it look as if it were much older than it really was. The skeletons in the rocks may look millions of years old, but that is because they were created to *look* as if they were old. Using the speed of light we may think the background radiation started billions of years ago, but God created it 6,000 years ago and put it rather far away.

Nothing can refute this "as if" game, but the date of origin is entirely arbitrary. We can say the universe began ten seconds ago. At that moment everything popped into existence, including our memories of a non-existent past. Note that Ussher did not dispute a realist position about the existence of the physical universe. It is real today: it just was not there before God created it.

To put this line of argument into perspective, assume the universe pops into existence every second. Also assume each successive universe is a duplicate of the preceding one looking as if it were one second older. I find this comforting. If I only knew that this succession would persist into the future, it would solve the problem of induction.

As science has progressed, it has introduced novelties that focus us on new experiences. This is a strong argument for scientific realism. If theories merely accounted for experiences we had already had, you might conclude their job was just to account for human experience. Often, though, they lead us to experiences we never imagined, and this seems to indicate that theories are describing a reality independent of the experiences to which it gives rise. One of the greatest surprises is how new theories give previous theories a more and more coherent unity, as if these new theories were getting closer and closer to a set of natural laws that govern the universe and have shaped its history from its inception.

Consider this analogy. We tire of piecing together the sketches of hikers and take aerial photographs of an area of the English countryside. It gives us a good geographical map in color but we notice something odd: farmers have planted out most of the area and there has been a mild drought. Some foliage shows up as dark green, some appears less green, and some is tinged with yellow. Curious, we dig and find Roman roads under the dark green patches and the foundations of Roman homes under the yellowish patches. This is a surprise: the area was not thought of as one of Roman settlement. Our map not only captured the experiences it was meant to—those of hikers—but also led to a whole new realm of unanticipated experience.

What is the explanation? Agricultural experts tell us that the sunken roads have filled up with topsoil and that therefore the depth of topsoil above them is thicker than average, and that the concrete foundations of the houses are close enough to the surface that the topsoil above them is thinner than average. Therefore, after a period of drought the crops above the former roads are the greenest and the crops above the former homes are struggling. The farmers had noticed differences in fertility but never understood why.

We wanted a simple map and that dictated our method of procedure, but the significance of what we got transcended that. Our new map and theory surprised us by unifying the geography, archeology and agriculture of the area. It was as if there were an underlying set of relationships waiting to be discovered. It led us to new unanticipated experiences, because there was a reality waiting to be experienced.

The history of science is all about new theories that not only better covered the known data, but also led to unexpected experiences and unification of earlier theories. The history of astronomy is a case in point.

Ptolemy (85–165) used geometry to account for virtually all observations available in his day. He made two mistakes: he put Earth, rather than the sun, at the center of the solar system, and he thought the planetary orbits had to be governed by circular motion. Accordingly, he had to use epicycles to get the right orbits. Each planet moved, not in a smooth curve, but as though it were on a Ferris wheel. The big wheel moved in a big circle and the planets were like passengers in chairs attached to the wheel, who moved around that point in their own little circles. As more precise observations came to hand it became clear that the planetary orbits could not be fully described in this cumbersome way.

Galileo (1564–1642) put the sun at the center of the solar system but kept circular motion. In reality, Earth not only moves around the sun but does so in a sort of squashed curve called an ellipse. The sun is not quite at the center of this curve but slightly to one side of the mid-point of its longer diameter. Therefore, Galileo had to say that Earth moved around an empty point in space, close to the sun but not identical with its actual position. This seemed very strange.

Johannes Kepler (1571–1630) broke the paradigm of circles. He hated to do this as he loved circles, but he had to accommodate new data on Mars, the orbit of which is even more squashed out of a circular shape than Earth's. He finally faced the awful truth: the orbit of Mars is a big ugly ellipse. Kepler did not just explain the known observations better, he introduced the concept of gravity, the notion that the heavenly bodies tugged on one another.

This led him to predict something novel: for a person on a spaceship going to the moon, there would be a point of zero gravity. At a certain point, the tug of Earth would be exactly balanced by the tug of the moon and the space traveler would be weightless and float about in the cabin. No one had ever imagined such an experience but today we know he was correct.

Isaac Newton (1643–1727) discovered the laws of motion that lay behind Kepler's map of the heavens. He showed that the solar system and its elliptical orbits, and indeed all of the heavens, made sense once you assumed that heavenly bodies attracted one another in proportion to their mass, and inversely to their distance from one another squared. His theory also brought a wonderful unification. Galileo had offered equations that fitted observed orbits of all objects moving about in Earth's gravitational field, such as a ball shot from a cannon. Galileo's equations were found to be just a special case of Newton's equations. Newton's theory had unified astronomy and mechanics (see box opposite).

As observations of the orbit of Mercury improved, Newton's predicted orbit did not quite fit. Einstein (1879–1955) explained everything Newton did, plus the orbit of Mercury, plus all sorts of new things such as the bending of light. He assumed that space and time affect one another, and that the presence of matter warps four-dimensional space-time into a curve.

This led scientists to anticipate many observations they had not previously contemplated. For example, if light travels along the line of least distance and if we live in curved space-time, then the angles of triangles in the heavens—the light that connects the sun and two planets—will add up to more than 180 degrees. Try it on the surface of a globe. You will

ROBERT HOOKE AND NEWTON

Robert Hooke (1605–1703) anticipated many of Newton's insights about the new universe entailed by the law of gravity, but he lacked the mathematics to elaborate the system. Annoyed at Hooke's claim to precedence, Newton denied him any credit at all.

Hooke built the apparatus that Boyle used for his experiments on gases. He was the first to use the term "cell" to describe biological organisms. He discovered one of the first double-star systems, Gamma Arietis. When Christopher Wren designed the dome of Saint Paul's Cathedral he used a method of construction that Hooke suggested. Hooke was also the first to advocate a grid pattern for city streets and avenues. You can see why he became rather cranky that his name was attached to so little. All he got was Hooke's law, which refers to the elasticity of springs.

find that the triangle you trace bulges at the angles. No one had ever thought of such a thing, but when we look at the heavens we have the experiences predicted (see box on next page). (Consider the absurdity of imagining a hypothetical observer: when we said a triangle within a distant galaxy had more than 180 degrees, we would mean that if there were a human there with a telescope, he would see it as such.)

Einstein wanted to discover a broader theory, one which would unify the equations that separately explained gravity and electromagnetism. James Clerk Maxwell (1831–1879) had already unified electricity and magnetism: he began

SPACE, CURVED AND FLAT

While we can talk about local curved space (near Earth, near the sun, near a star in another galaxy) the total space in the universe may be flat. This is called the "Einstein–de Sitter universe". Imagine a huge blanket hovering flat in mid-air. It is stiff enough to stay flat but elastic enough to give if something rests on it. If one ball lies on top of the blanket, it will curve the whole into a downward slope. If there are balls (or cubes) on every inch of it, the "downward pressure" will be uniform and the blanket will still be perfectly flat. If there are many balls separated but approximately evenly spaced, there will be local downward slopes but the overall downward pressure will still be uniform. If you walked across the blanket you would pass through many "valleys" but you would be at the same elevation at the end as when you started. As to the shape of the universe as a whole, that is determined by whether its total matter/energy is greater than, equal to or less than a critical value at which gravity would be just sufficient to counteract the expansion set off by the Big Bang (plus the effect of dark energy if it exists).

with twenty-five equations and condensed them into eight. (Later they were condensed into four, although you actually need a fifth, which Maxwell had included in the set of eight.) Einstein was frustrated by quantum theory, which deals with the behavior of particles within the atom. To his dying day he attempted to find elegant geometrical laws that would explain all the forces of nature. He failed.

Today new theories, as yet untested, unify all four fundamental forces of nature: gravity, electromagnetism, and two within the atom called strong force (which holds the nucleus together) and weak force (which governs the exchange of fundamental particles). String theory does this perhaps most successfully in a version that conjectures that space-time has eleven dimensions. We are able to observe only four of these, the familiar three dimensions of space and one of time. The others are too "flat" for us to see.

It makes sense that we are creatures who cannot see every dimension. Imagine you are a flat creature that lives, like a postage stamp, on the surface of a flat plane. You will see only two dimensions in any direction. The plane may, however, exist in a three-dimensional world that affects you. If a sphere passes through the plane, you will first see a dot, then a circle that gets bigger, and then the circle dwindling away to a dot. To explain what you see, you will have to invent a theory that takes into account the unseen third dimension.

Thanks to evolution we can observe only a small portion of the universe, namely that part of it we need to see in order to survive. Our ability to see Jupiter is an accident of evolution, but Jupiter's existence is not an accident of evolution. Jupiter is part of the universe at a certain stage in its development.

Physicist Jacques Distler (born 1961) has derived from string theory precise predictions about collisions between subatomic particles. The new Large Hadron Collider near

Geneva may supply the enormous energy needed to detect them. Raphael Bousso (born 1971), another string theorist, has expanded on the idea that our universe is only one among an infinite, ever-growing assemblage of disconnected bubble universes. Using that hypothesis, he has derived an estimate for dark energy in our own universe that closely matches observations. His work holds out the prospect of a multi-universe theory that will generate novel predictions for our own bubble universe. It may even generate calculations that apply to the multi-universe—the collection of all possible universes—even though the other universes are unobservable, lurking beyond our cosmic horizon.

There is no guarantee that we can collect all the observations needed, or are bright enough to invent the theory needed, to fully explain the physical universe or universes. We may have to be content with getting closer and closer. But the idea that science is all about explaining what we happen to experience is bad history. The job of science is to account for our current experiences, certainly, but the better the job it does of that the more it stumbles on new experiences and unanticipated explanatory power. It emerges as a vehicle for picturing a real world beyond human experience, structured by laws of nature that impose order on the whole (see box opposite).

We can now assess the thinkers who attempted to make sense of the rise of science. Thomas Hobbes was right to think that scientific concepts improve our understanding of the

LAWS OF NATURE
You may feel outraged that I have spoken about laws of nature. Doesn't the problem of induction make nonsense of the idea that laws of nature bind the future to the past? It does indeed, but remember we have agreed to set that particular problem aside. Once we do so there is a strong case for such laws.

physical universe. Ironically, however, he took a step that eventually obscured this insight by calling the molecular structures of the material world the only real entities, and the world as we experience it mere appearance.

The universe constantly evolves to produce new things, all as real as the things that preceded them. A universe of elementary particles became a universe of matter and galaxies, which evolved planetary systems, one of which (at least) evolved life. Some of these life forms evolved consciousness, and some with consciousness evolved a sense of personal identity and the kind of reflections about experience that we call science. Even though they did not always exist, my experiences and thoughts are just as much part of the universe as is any other part.

Hobbes' assumption that only bodies in motion are real led him into reductionism. He believed, for example, that he had to explain forgetting in terms of impressions rippling through the "atoms" of our mind and eventually fading out.

Treating all things as real suggests we need many levels of explanation. We may someday be able to "see" forgetting occurring in the brain by magnetic imaging of its neurons, but physiological explanations will never render superfluous psychological explanations (why you always forget your partner's birthday) or sociological explanations (why you do not kill your opponents to win at games).

It is a wonderful achievement when science uses one level of explanation to elucidate what is going on at a "higher" level. Linus Pauling (1901–1994) reduced the chemical table of elements to physics. Using quantum mechanics, he showed how molecular bonds arise from the properties of the atoms of which molecules are composed.

But this did not eliminate chemistry as a discipline. We still experiment to see what new compounds we can produce in the laboratory. In theory, we might predict the movements of every woman in Boston from brain physiology but this would not tell us why, at one time, most prostitutes were Anglicans. William Graham Sumner (1840–1910), the great pioneer sociologist, found that most prostitutes had been in orphanages and that Boston orphanages were all run by the Anglican Church. We may be composed of elementary particles but they do not live our lives for us.

Higher levels can set problems for lower levels. My work on intelligence suggests that the brain differences that distinguish individuals from one another today differ from the brain differences that distinguish us from our ancestors. Identifying the two on the level of neurons sets a new task for brain physiology. And science on one level can correct mistakes we tend to make on another. Human genetics has changed the view that social environment accounts for most of the differences between the intelligence of individuals. On a psychological level we feel as if we have free will, but that belief may be contradicted by findings on the physiological level.

John Locke said that material objects have position, density, shape, number, and sometimes movement. Today we can say much more about them of course, that they are composed of cells, molecules, atoms and an array of subatomic particles. And we know that matter, as Locke thought of it, did not exist in the early universe. As for objects' other properties, he distinguished effects—such as color—that arise when objects interact with our sense organs from those that arise when objects interact with other objects—for example, fire melting wax. He did not imply the former took priority over the latter.

Locke didn't doubt we could "model" what the physical universe would be like without us. The universe in that state was no less real than the universe we began to experience after we came to exist. He was a realist; he bears no responsibility for the confusion that followed.

Kant was wrong to think space and time are products of the spectacles through which we see the material universe. We see straight-line space even when space-time is curved, therefore his assumption that the geometry of Euclid would describe the phenomenal universe was mistaken. We cannot salvage Kant by saying that we are just learning more about the structure of the spectacles. He got the spectacles right:

physiology confirms that we see the world as he said. His mistake was to deny that the physical universe is oriented in space independently of experience.

Kant affords an opportunity for an aside. What existential status do mathematics and the "laws" of nature possess? Rather than debating whether they exist, it is better to describe *how* they exist. The rules of logic and logically coherent mathematical systems do not exist like tables and chairs that can affect us whether we are aware of them or not. (I sometimes bump into tables in the dark.) They exist as permanent possibilities, awaiting the emergence of minds capable of logical reasoning. If such minds exist on other planets and think about Euclidean space, they will derive the same geometry we learn at school.

As for the laws of nature, they exist as possibilities that await their appropriate subject matter. The laws of chemistry could not bite until entities with a molecular structure emerged, the laws of evolution could not bite until there were living things, and the laws of human psychology could not bite until humans evolved.

The logical positivists were mistaken when they ruled that saying something exists means we must be able to imagine experiences to test its existence. Like Bertrand Russell, I can imagine no such test of God's existence, yet I understand perfectly well that when a priest says "God exists" he means that God exists independently of the physical universe, just as the physical universe exists independently of me.

This brings us to the philosophy of "ordinary language". To appeal to how people speak is absurd unless they speak as rational people ought to speak. People may talk about God, but if God does not exist this is as silly as talking about astrology when we know that the stars do not really influence our lives.

Theology is not a language game with its own "logic". It may pose "why" questions about human existence, calling for answers beyond those that science can supply, but no sincere religious believer will say God exists simply because we could satisfy certain yearnings if he did. That is no more respectable than saying Santa Claus exists because he satisfies the logic of Christmas talk, providing an answer for children who think their presents have some extra-parental source. Religious people believe that God exists literally, and that he would exist even if not one person yearned for such answers or spoke a language that posed them.

Kuhn was correct when he said that throughout its history science has shown paradigm shifts. Indeed what Kuhn said about paradigm shifts Einstein anticipated. Einstein stressed that as long as Newton's world view held the field unopposed, evidence such as minor discrepancies in the orbit of Mercury would be ignored because they seemed to make no sense. For science to progress there had to be an opposing world view, namely his, in which the orbit of Mercury made sense.

Sometimes the need for a new paradigm is foreshadowed by the fact that just too much data is accumulating against the old theory. However, the fact the data cannot be explained without a radically new theory does nothing to discredit realism. Each new theory may be a better approximation of the laws that govern the universe.

I have said enough about Berkeley and van Fraassen. The fact we can learn about the physical universe only through experience does not mean we can learn only about our experiences. Once we accept naïve realism, Berkeley's idealist spin on realism and van Fraassen's limits on scientific realism are both untenable.

Every day, both life and science teach us more about a physical universe of which our experiences are just a small part. It is a universe that existed long before we did and will persist long after we are gone. We are real only because it is real.

DOES GOD EXIST?

The physical universe exists but does anything else exist? Are there entities that transcend the world of space and time? God is the transcendental entity for whom the most persistent claims have been made. These claims are so prominent that every person is our society lives as if they were either true or false.

There are two kinds of case for God's existence: proofs and appeals to faith. Saint Anselm (1033–1109) offered a proof in three steps. Our concept of God is of that greater than which nothing can be conceived. What exists actually and mentally is greater than that which exists mentally alone. Therefore, we must conceive of God as existing both actually and mentally. Failure to do so, he argued, involves a logical contradiction. If you conceive of God as existing mentally alone, you can conceive of him as existing actually. And then you will have conceived of this: something that was greater than a thing greater than which nothing can be conceived.

This prompts an analysis of two of Anselm's assumptions. Can the concept of a thing be said to be greater or lesser than the thing itself? And does conceiving of something as existing alter our concept of the thing?

Take a photograph of an elephant. The photograph is not a small or diminished elephant: it is not an elephant at all. How are the two comparable in terms of "greatness"?

Indeed, how are they comparable at all? I judge one in terms of the sharpness of the image, how much of the elephant is visible, and so on. I judge the other in terms of how large the elephant is, how strong, how healthy. The elephant can trample me and the photograph cannot. But that is merely to say that one is an existing elephant while the other is merely an image of an elephant captured on film. One does not exist more than the other: one elephant exists and the other "elephant" does not exist.

Replace "photograph of an elephant" with "conception of an elephant". The conception of an elephant is not comparable to a real elephant in terms of the former being a lesser elephant and latter a greater elephant. Nor are they comparable on some strange scale of existing more or less.

This raises the question of whether conceiving of something as existing alters my concept of the thing. At this moment, I am conceiving of both my father and my son. Both images are equally sharp. I believe that my father no longer exists (he died in 1955) and my son still does. But if I were mistaken, if my father rose from the dead and my son died ten minutes ago, my conceptions of them would not alter at all. My beliefs about whether or not they existed would alter—I could conceive of one or the other as a corpse—but these beliefs are no part of the conceptions. My father and my father's corpse are not the same person. A corpse is not a person at all: it is simply a cadaver.

We are now aware of the hidden assumptions of Anselm's argument—namely, that conceptions of things and actual things are comparable as to greater and lesser, and that existence is an attribute of our concept of a thing. We have rejected both of these assumptions.

With this sophistication, a wary atheist will want to restate Anselm's propositions as follows. First, we do not all "conceive" of God in the same way. We all share a concept of an entity that is all-powerful, all-knowing, and benevolent, but some of us accompany that with a belief in his existence and others with no such belief.

Second, to say that something that exists both actually and mentally is greater than something that exists mentally alone is absurd. There is no common metric between a thing and the concept of a thing. Third, I never conceive of anything as existing both actually and mentally. I conceive of a thing and then either add, or fail to add, a belief that it exists.

Therefore, Anselm's proof comes down to this: if you accompany the concept of God with a belief that God exists, it is contradictory for *you* to deny that God exists.

Thomas Aquinas rejected Anselm's proof in favor of five arguments of his own. We cannot, he said, posit an infinite regress of causes of what exists going back into the past, nor an infinite regress of causes of motion going back into the past, therefore we must presume there is at least one uncaused first cause or unmoved prime mover.

His third argument was that the universe cannot be composed entirely of contingent beings. Contingent beings are things whose existence is not guaranteed but dependent on certain circumstances. For example, I will exist only as long as my heart keeps beating; our planet will exist only until the sun expands and consumes it. With an infinite amount of time going into the past, however high the odds against every contingent thing's number coming up at once, it would have happened. Think of each contingent thing's existence at a particular moment being dependent on someone not getting ten heads in a row when they flip a coin. No matter how many contingent things there are in the universe, given an infinite amount of time all of the coin-flippers would get ten heads in a row at the same moment.

If that had ever happened in the past, nothing would have existed then, and therefore nothing would exist today. Hence, there must be at least one being whose existence is not contingent, a necessary being that needed no cause to bring it into existence, an uncaused first cause. Note how this proof depends on setting aside the problem of spontaneous occurrences: if things can pop into existence out of nothing, things could exist today despite nothing having existed yesterday. However, we earlier agreed that all schools of metaphysics had a license to set that problem aside.

Aquinas's fourth argument is that we all make comparisons of value—more or less beautiful, more or less righteous, and so forth—and so must have in mind some absolute standard not of this world. This argument has attracted few adherents so I will dismiss it quickly. The obvious answer is that even if an absolute standard is necessary to make these comparisons, if we use it then it must be in our heads. There seems no reason to assume it also exists outside our heads (like some sort of moral reality).

The fifth argument stems from the concept of design. When we look at the physical universe, not only the heavens but also the wonderful variety of living creatures, we see something that presupposes a purposeful creator with a rational plan for the universe. As William Paley (1743–1805) once said, if you see a watch on the heath, you assume there is a watchmaker.

The classical objections to these arguments are persuasive. First, an endless succession of causes into the past is no less sensible than an endless succession into the future. At any time in the past you want to cite, we can point to what began a series of causes leading up to present and that is good enough.

Second, any particular thing in the universe is contingent only in the sense that it depends on something else to keep its present identity. It is true that if my heart stopped beating—for example, because it was without oxygen—I would die and decompose. But if I did die I would merely sink into a universe, the existence of which is not at risk even if every entity in it decomposes or disperses. If I must deem something to be a necessary being, I offer the universe itself.

Third, the theory of evolution gives an explanation of the apparent design of living things without assuming the existence of a watchmaker. The Hawk moth caterpillar has a rear end that looks exactly like the head of a snake. That means it is less likely to be eaten by birds that are frightened of snakes. As for intelligent design, if you believe it adds to God's dignity to say he wanted the rear end of a caterpillar to look like a snake, that is up to you.

Some of these arguments have been revived by recent developments in cosmology—theories about the origin and development of the universe. Perhaps the universe is not a necessary being. We now think its existence was contingent on something else, the singularity that led to the Big Bang. Over thirteen billion years ago there was an incredibly small package that exploded to evolve into the universe of today. The singularity acted as a first cause.

Moreover, it had a design: if the package had contained laws of nature that differed by the slightest degree from what they are, and if the infant universe had lacked tiny irregularities that saved it from total "smoothness", we would have had a universe about as interesting as an ever-expanding bowl of oatmeal. From this, it is argued that an intelligence that transcends the physical universe must have added an element of design.

Let us concede that all this is true. Of what possible interest would this remote mind, which prefers some universes over others, have for a religious person? There is nothing in the Big Bang concept of God that demonstrates an interest in me, or indeed any special interest in our universe. Thirteen billion years ago God may have created other universes he found even more interesting. I certainly have no reason to believe that he would be interested in our universe because it contains people. People were not around for almost thirteen billion years and will not be around for endless billions of future years. We may be a blot on the landscape compared to the delights of spiral galaxies.

For this god to be of interest, he would have to give me a revelation along these lines: "The Big Bang is not just the latest scientific theory but the truth. I am the mind behind the singularity and I designed it to produce human beings. I take a special interest in them. If you obey the following commandments, I will give you eternal life." If I got such a revelation I would be far more interested in it than the cosmology. Without it, the cosmology is only of impersonal scientific interest.

Most people, though, do not base religion on proofs. They base it on faith. Here we must distinguish true faith from conventional belief. If your religious faith is merely part of the baggage you were raised to believe—that New Zealand is the best country in which to raise children, that America is the only free nation, that your father can lick any father on the block—I leave you to assess how much it is worth. The only faith that has dignity is a personal relationship with God based on a sense of his presence sustained through contemplation and spontaneous prayer. I will take someone's faith seriously only if they spend some time each day in non-ritualized prayer.

Even here we must ask whether the intensity of true faith vouches for the believed-in god as real, rather than fictitious. The usual answer is that if you had true faith you would never doubt it. Until the age of ten I had absolute faith, inclusive of sincere prayer. I lost it when I realized that I believed in that particular faith only because those who reared me had it rather than another faith. The communion with God was deeply comforting, but upon reflection not distinguishable from "communication" with an imaginary friend (see box opposite).

The usual answer to this is that if I had really had true faith, I would not have lost it. This is an empty tautology: true faith is ever-persistent faith; therefore, no lost faith counts as true faith. Anything can be defended by definitions that make it safe from being falsified by evidence. All blacks are stupid, so any black that appears to be an exception just must have some white blood.

Those who establish a personal relationship with a god are likely to be bathed in a particular religious tradition. They find it natural to take the depth of this relationship as a voucher for the tenets of their religion, whether the real presence of Christ in blood and wine, or circumcision, or a paradise with dark-eyed virgins in their tents. There is no more sense in this than in taking your relationship with God as a vindication of your views on art or music or horse breeding—unless, of course, you know that the quality of your relationship with God is superior to any on offer

RELIGION AND ETHICS

You may say, why not abandon humane ideals once you realize you hold them only because those who reared you held them? The answer is two-fold.

First, to postulate that something exists only because you have a strong inner need for it to exist ignores the fact that things just exist or do not exist: our desires have nothing to do with it. To claim that ideals are precious because you are strongly committed to them is not subject to the same absurdity. Ideals exist insofar as people are committed to them and that is the end of the story.

Second, you may not hold your ideals simply because one or both of your parents held them. My father and mother both had mild racial prejudices, and my father a more than mild hatred for the English crown. As I grew up I found that these were foreign to me. When I read about blacks and Jews in literature, and later met them, they seemed just as valid people as Irish-Americans.

My peers, too, were racially biased but I found that was because they had false stereotypes. Almost everyone I knew was a "patriot". As I learned about the actual behavior of nations, including America, I decided I was not a patriot. I had to grit my teeth and judge America as impartially as I would any other nation.

I do not deny there must have been deep inside me a core of humane ideals with cultural roots somewhere, but through personal experience and reason I "purified" them into moral ideals that defined my own distinctive deep inner self.

from Islam or Orthodox Judaism or Jainism, but this is a judgment no one is in a position to make. To say that you have faith that your relationship is superior is to pile one absurdity on another.

To vindicate your particular religion, God would have to send you personal messages telling you that Christ is the redeemer, or that Mohammed is his prophet, or that the laws of Moses still prevail. For a Christian, such a message would not be nearly enough. God would have to read you a long theological tract that attested to the virgin birth and the resurrection, detailed the code by which one must live to attain salvation, and described the legions of angels and devils. I have never met anyone who claimed to have received such messages, nor did I get them when I had absolute faith.

All religions offer me hearsay evidence that their prophets have received messages of the required sort, but none of the evidence seems remotely reliable. In addition, such claims outrage my sense of justice. That some would have direct evidence of the truth and others must accept it on faith—not in God, but in the veracity of a person—seems to be the worst kind of discrimination. Certainly God would have communicated his truths to us all. To make most of us dependent on the word of fallible men or women seems obnoxious, especially since so many who have claimed to have received such messages have later been denounced as frauds by their followers.

If you take the divinity of Christ seriously, you owe it to yourself to read the literature, not just the Christian apologetics and the skeptics but also the Jewish historians. My own reading is that Christ did not claim to be God or even the non-divine Jewish Messiah, but rather thought of himself as a Jewish holy man with the same message as his mentor John the Baptist. He had come to warn the Jews that the coming of the messiah was near and that they were unprepared. They thought obedience to the laws of Moses was enough. They were mistaken: they needed to have charity in their hearts and manifest it in their lives.

I base this on a reading of the gospels, the Acts of the Apostles, James the just, and Saint Paul. The older gospels, Matthew and Mark, appear to be a mix of genuine stories about Christ that were edited by Paul's converts after they came to dominate Christianity. The problem was that these were sincere people who did not think of themselves as editors, and therefore anecdotes survive that suggest a Christ unlike the god/man of Christianity.

When a gentile woman approaches Christ to heal her son, he says gentiles are dogs. He relents when she pleads that even dogs can beg at the table, but the point is that Christ does not seem to think his message is for non-Jews. James, who knew Christ and succeeded him in the sense that he headed the Christian synagogue in Jerusalem, barely tolerated Paul's attempts to convert non-Jews. Paul, who did not know Christ, claimed to have received a revelation from God as to the divinity of Christ. The letter of James makes no such claim and does not mention the resurrection or the virgin birth.

Mary's behavior as reported in the gospels sometimes seems inexplicable. When the twelve-year-old Christ preaches at the temple, Mary leaves Jerusalem without him and only later notices his absence. When she remonstrates with him, he has to tell her he has been about his father's business. Her negligence and state of mind seem odd in a mother who, every day of her life, was aware her son was divine and of miraculous birth.

I do not know that Christ thought of himself only as a Jewish holy man, I merely think it probable. I can understand others concluding that a claim to divinity is more probable, but anyone who tells you they are certain of such a claim is being fatuous. No one can be sure of Christ's psychology, any more than we can be sure of Abraham Lincoln's psychology. Indeed we have far better evidence of the latter.

Setting aside whether a claim to divinity can be taken at face value, the Christian faith rests on probability rather than certainty, and looming over all is the specter of a god giving tantalizing, rather than clear, evidence of his divinity, as if he did not mind alienating many of the most honest and sincere would-be believers. He also seems unconcerned about whether people suffer through no fault of their own (see box opposite).

French philosopher Blaise Pascal (1623–1662) proposed what has come to be known as Pascal's wager. Even if the chances that Christianity is true are slight, he said, look at what is at stake: eternal bliss if I follow its tenets, eternal

THE CHRISTCHURCH EARTHQUAKE

In 2010, an earthquake in Christchurch, New Zealand killed almost 200 people. A churchman said the earthquake dispelled complacency and brought the community together. If it had turned the community into a pack of dogs tearing one another apart, he would still have found some good in it. If it had killed only innocent infants, there would be a rationalization. After all, every orthodox Jew and Christian has already rationalized such a crime: God killed all the first-born of Egypt to punish the Pharaoh (Exodus 12: 29).

In the Old Testament, Jehovah is a tribal god for whom the lives of non-Jews simply do not count. Some say the god of the New Testament is different, as if God had entered some kind of clinic and emerged reformed. Others, made of sterner stuff, await the day when an angel of God will send a plague of locusts with faces of men, hair of women, teeth of lions and tails of scorpions to torture those "which have not the seal of God in their foreheads" (Revelations 9: 1–10).

Christians rationalize evil in two ways. First, that it is not very evil: Anglican priest Gordon Graham says the fact we do not all commit suicide is a tacit endorsement of God's handiwork. Second, that it is good in some way we cannot perceive.

The final step is to eliminate God's divinity by turning God into an ideal of what is best in man. This makes the "believer" identical to the atheist. Most atheists cannot understand why a few atheists want to use language (talk about "God", for example) that makes it difficult to convey what they actually believe.

damnation if I do not. If it is false I lose little by obedience, just keeping the commandments, which I might want to do anyway, and time spent in worship. If it is true I lose everything by my defiance. It makes sense, then, to bet on its truth. It is like being given a free sweepstake ticket that offers me some chance of bliss and absolute insurance against the greatest possible disaster.

Pascal did not deny that you would need true faith in order to be saved. However, he held out the prospect that if, thanks to his wager, you lived as if you believed, that way of life might engender true faith.

The problem with the wager is that it in no way dictates gambling on Christianity in particular. If another religion that demands my obedience is true, Christianity leads to eternal damnation. In theory, one could experiment by giving the life every religion recommends a try and seeing if one is unique in engendering true faith, but life is too short for that. And surely it is antithetical to a religious person to try different religions in the order in which they appear on a balance sheet: which one promises the most and asks the least. If we really want a good bet, Christianity cannot compete with worship of the Great Purple Tree Hog: all the latter demands is saying the tree hog's name once a day. In return he promises to match the best heaven on offer. If you think this religion is less sensible than other religions, try to prove it.

Pascal's wager captures the kind of fear that often sustains religious belief. I refer to the famous maxim "There are no atheists in foxholes." When frightened of death, almost everyone can at least understand asking God for help. This plea—"I do believe so please help me"—is an inarticulate version of the wager. It really says: "I believe because belief may save me from a terrible fate and non-belief leave me without hope." It is preferable to make up your mind about what is true or false when your reason is functioning normally, not when it is overwhelmed by fear.

One last point: when I analyzed the problem of induction, I said that every theory of being, whether scientific realism or Christianity, had to set aside the possibility of spontaneous occurrences. Since no rationale can be provided for doing this, it is an act of faith of the most arbitrary sort. If totally arbitrary acts of faith are legitimate, why must religious belief be based on what I call true faith? Why should someone not just flatly say he or she has faith that the bible is the unadulterated work of God without further ado? It appears we all have a blank check, which we can fill in with any kind of faith whatsoever. It is true that the check is meaningless until it is filled in—it signifies faith in nothing particular— but once you have filled it in for Christianity you can hardly fill it in for Islam or Buddhism or anything else. You have a secure Christian faith and that is that.

My response is that this shows amnesia concerning the peculiar character of the "act of faith" that dismissed the possibility there would be spontaneous occurrences. We were driven to assume something that we knew could be true or false. This is hardly a prototype for faith in God, that we believe in him but with the reservation that his existence may be fact or fiction.

The second problem with blank-check faith is that it indiscriminately allows subjective certainty to qualify as faith. Faith is reduced to the lowest common denominator of sheer belief with no questions asked. The only thing that would flunk would be conscious hypocrisy. The fleeting faith of a troubled person becomes the equal of Saint Augustine's matured faith. Any sincere person can claim faith in anything under the sun. You can raise no objection if your daughter comes home worshipping a foreigner who demands his followers match his weight in oatmeal.

What has happened to the claim that Christians have faith in what truly deserves to be believed? It has dissolved into a "right" of everyone to claim that his or her subjective certainty is faith, even though we know from experience that subjective certainty is often misleading. I was once certain that Ted Williams had achieved a batting average of exactly .400 in 1940. A look at the record book showed I was mistaken.

Atheists are not the only ones who want faith subjected to an analysis of what it validates. Any religion that seeks converts must ask people to question whether their current faith vouches for what they think it does. Any thinking person who has an intense relationship with what they regard as a transcendental being wants to subject their own faith to analysis concerning its significance. I stand by my thesis: religious belief is either based on a personal relationship with God or based on nothing worthy of regard. And even if we give true faith the benefit of the doubt, analysis of that relationship shows that it justifies no belief other than the existence of God. This is to say, true faith, properly understood, leads straight to deism.

Deism, strictly speaking, is the belief that God created the universe but takes no part in it. It rejects all conventional dogmas and false history in favor of pure worship of the deity. It promises no rewards or afterlife and imposes no obligations. When a deist invites me to share what he or she experiences, the motive is benevolence. It is like telling someone they are missing out on a wonderful friendship, an enormously satisfying relationship pursued for its own sake. But if I am happy with what my life offers me, my work, my loves, my present friends and my sport, and have difficulty taking seriously the existence of this new friend, I see no reason to try to cultivate belief—particularly since I know from my own childhood what is on offer, something as nice as a wonderful daydream but no more.

DOES THE ABSOLUTE EXIST?

This chapter completes my discussion of what I think exists and does not exist. By its end, all of the problems that kept me awake at night during some period of my life will have been "solved". Today is my seventy-seventh birthday, so it has taken me sixty-five years to replace Catholicism with a personal philosophy I can live with. This book is intended to give you a head start.

The mystical experience is sometimes classed as religious and, that aside, has its attractions. My account relies heavily on two great students of mysticism, William James (1842–1910) and Walter T. Stace (1886–1967).

When some people empty their normal consciousness of all its usual content—sense experience, concepts and emotions—rather than their mind remaining blank they are aware of an overwhelming consciousness. Often called the Absolute, this is accompanied by a quiet all-absorbing bliss, a sense of liberation from the strictures of self and time and space.

There are various methods of attaining this state but all are techniques for ridding the mind of its normal content. Meditation on a single object—for example, on the halo of the Virgin Mary—if intense enough, kills awareness of all other objects. Then, because we can be aware of one object only in contrast to another (even if only in contrast to a background), that object fades and the consciousness is blank. This is captured in the Zen kōan, "What is the sound of one hand clapping?" The answer, of course, is total silence.

A Zen master may use a paradoxical remark, or an unexpected tap on the cheek, to shock a novice out of normal consciousness. An economist once told me that the trauma of someone trying to strangle him transformed his consciousness. For a few, flagellation may be necessary to quiet a sense of guilt that nags at the edge of consciousness and prevents them from losing awareness of self.

It will be apparent from this that the mystical experience is not a religious vision—for example of the Virgin Mary, or Jesus walking on water. These have the same content as a sense experience. In the mystical experience all such content is absent.

Normally our experiences are divided into two sorts: either they have a specific content and are an awareness of something beyond ourselves, such as when we hear a note of a particular pitch and volume (sense experience), or they have a general content and are within ourselves, such as when we think about the notion of a category (conceptual experience). The Absolute has a general content akin to a concept and yet is an awareness of something that transcends the self, akin to a perception. It is as if one could "see" the concept of a category.

It is not an emotion. It is accompanied by beatitude, which is like an emotion in that it is a sense of liberation, but does not touch the core experience of apprehending something, just as seeing an old friend may excite joy without the image of our friend being altered. The feeling of beatitude results, at least in part, from the sense of transcending time and space. The Absolute is not here rather than there: it is all. It is changeless so there are no events happening over time. Although an apprehension of an other-than self, it is isolated from the five senses. The Absolute says nothing about the realm of time and space and they say nothing about its timeless realm. It is the sole "sense" that explores its realm.

There are two variants of the mystical experience. One is the introverted experience I have described up to now, in which the Absolute replaces the inner self. The other is the extroverted experience, sometimes called nature mysticism, in which the Absolute "replaces" the external physical world. You look at a bucolic scene in which a horse is grazing by a stream near a grove of trees. Suddenly the scene is transformed. The horse, stream and trees are still there, but it is as if they were translucent and their specificity missing, as if they had been changed into a translucent checkerboard with the Absolute shining through all the squares equally, robbing them of their particularity, including the particularity of their position. The squares are interchangeable in a way that robs position of relevance. The scene is frozen and thus eventless.

Only a few mystics have had both experiences, the most notable being Meister Eckhart (1260–1327), a great introverted mystic, who also reported that he saw God shining at him through a leaf.

The writer Aldous Huxley (1894–1963) believed that certain drugs help attain at least an approach to the extroverted experience. This is an empirical question, but adherents of the mainstream mystical tradition believe he was mistaken. They liken the effects of drugs such as LSD to Walt Disney's *Fantasia*, and warn that taking such drugs is likely to become an end in itself and impede attaining true mystical awareness.

Mysticism gets a bad name because the mystics themselves say that the experience is ineffable and cannot be described in words, or can be described only in paradoxical language. This is understandable. If, for example, I were speaking with a woman who was blind from birth and I wanted to get across what it was like to take in a huge expanse of calm water at a glance, I might say, "It is like touching a smooth sheet of marble many times at once." That is, of course, a paradox but still a good way of describing what I can see and she cannot. And in this case I have the advantage of speaking about a world that can be explored by both touch and vision, rather than by a single sense.

Much of what seems paradoxical about mysticism seems less so if you bother to learn something about it. We have already looked at the sound of one hand clapping. Take a phrase such as "Atman (my psyche) is Brahmin (the Absolute psyche) but is not Brahmin". Remember that all spatial orientation dissolves in the mystical experience. Therefore, you cannot be apart from the Absolute but apprehend it by having it become your consciousness: the distinction between subject and object disappears. On the other hand, the Absolute consciousness is qualitatively different from your normal self-aware personal consciousness. You are it but not it.

The great mystics of history are not the only ones who have had the experience. A relative once told me that when lying in a hospital bed his mind wandered and he had an extraordinary experience. It was as though a star suddenly burst into his mind. I asked the obvious question: did he literally mean he had the image of a star? If so, it would not have been a mystical experience. He said that no, he had not seen anything. "Star" was just as close as he could come to describing it. "It was as if I had suddenly become a ray on a star."

A student, after hearing me lecture, told me she had often had her "wonder time". She would be rocking rhythmically, vacating her mind, and her mind would dissolve, like a drop falling into an oceanic mind. Excellent words: a drop is not apart from the ocean but is not identical to the ocean. The great mystics, when under suspicion of blasphemy for claiming they had literally become God, used similar language to defend themselves. The Sufis were less cautious. They ran through the streets of Mecca shouting, "There is nothing under this cloak save Allah." They were often massacred.

CRUSADING MYSTICS
The greatest of the Spanish mystics were Saint Theresa of Avila (1515–1582) and Saint John of the Cross (1542–1591). Neither retreated from the world to simply bask in the mystical experience. Both were active in monastic reform at some risk to themselves. They demonstrate that one need not be a particularly unusual person to have the mystical experience. John writes about it in a way that shows high intelligence but Theresa is simply passionate about being the bride of Christ.

Since the Absolute is an undifferentiated unity, without dichotomies of any sort, it conveys no information about good versus evil. However, there are claims that the mystical experience is beneficial psychologically—that it automatically makes you a better person, for example. Certainly, mystics do not always lead a purely contemplative life. Sometimes they try to improve the lives of people in general (see box above).

I doubt we have enough data to arrive at firm conclusions about the psychological effects of mysticism. Most of the mystics of recorded history were religious people and so were predisposed to humane ideals. Finding that they and the whole physical universe were united in the Absolute often gave them a reverence for other people, nature and all living things. It appears the mystical experience acted as a sort of catalyst: their characters suddenly jelled in a way that might have taken years of virtuous living to achieve.

But what if someone with anti-humane proclivities learned how to empty his mind and escaped the bounds of space and time? Nietzsche might overnight have become even more convinced he was a superman, and justified in his contempt for ordinary people.

The mystical experience may work as a catalyst, fast-forwarding your character whatever your bent. This does not in any way diminish its value for the humane. An extraordinary sense of unity with an Absolute consciousness, a sense of being blessed, a quicker path to virtue—the combination makes a good enough reason to seek out a Zen master.

Does the Absolute exist? There are three possibilities. It is an existing entity, an addition to the objects of the physical universe (object possibility). It is something that shows the physical universe from a fresh point of view and reveals attributes we normally miss (attribute possibility). Or it is a kind of consciousness that rarely breaks through to awareness (psychological possibility).

The first possibility assumes a disembodied mental entity unlike anything in the physical universe. Even if this were the case, the situation would have no implications beyond deism. We might venerate the Absolute because of its splendor, but it would not offer a code by which to live in order to attain "salvation". No afterlife would be promised. The fact the Absolute would persist after you died does not mean you would still be around to experience it.

The second alternative is pantheism. A sphere looks convex from without and concave from within. Similarly, the physical universe may appear multiple and material when viewed from outside through the senses, but one and immaterial when, thanks to the mystical consciousness, viewed from within. Doesn't the extroverted experience show the Absolute shining through physical objects? According to the pantheist, everything has interior, mental elements. However, only some things—people and not rocks—are organized in a way that renders them functional as a conscious mind. In the same way, we have electrical elements throughout our bodies, but unlike those of the electric eel they are not functionally organized so we can give shocks. Perhaps the mystical experience reveals an attribute of the physical universe that we normally miss: its mental dimension when viewed from within.

Some claim the mystical experience is self-authenticating because the mystic has a sense of absolute certainty. But absolute certainty about what? Mystics have remained believers, pantheists or atheists, so a sense of "not doubting the authenticity of the experience" cannot discriminate between the three possibilities.

The usual case for the experience referring to something other than itself, either an object or an attribute, is based on unanimity. Whenever and wherever the mystical experience is reported, if you strip away the theology—such as the Absolute being called Brahmin, Jehovah, Allah or God—it is always the same. The only great religion not to report it is the Zoroastrianism of Persia, which does not mean that no Persian ever had it.

Normally, unanimity does not show that an experience is trustworthy. There are sometimes mass hallucinations, psychological states that can be induced in everyone using the same stimulus. Therefore, we need tests such as prediction, and cross-checking between the senses. If a group of people see an oasis 100 meters ahead, they will be upset if, when they reach it, they get a mouthful of sand.

The rejoinder is that these tests are reasonable when you are dealing with a realm which is explored by five senses and in which events occur over time, but inappropriate when dealing with a timeless realm explored by a single "sense". Assume for the moment that the Absolute does exist. How else could it reveal its presence except through unanimity?

This argument loses its force when it is realized that the unanimity is self-defined. We have a criterion of what qualifies as a visual experience: it is an experience affording images that have shape, color and position. The content of the experience is open. When two of us see the same thing, this adds information. We could agree that we saw a white creature rushing past, and disagree as to whether it was a dog or a cat.

However, an experience qualifies as a mystical experience only because it has the same content as other experiences counted as mystical. If it were not an awareness of an oceanic mind but instead a vision of the Virgin Mary, we would not call it mystical. The unanimity of the experience is not very impressive: angry experiences are classified together because of similar content but they do not vouch for an existing entity.

The foregoing reveals the key to deciding the significance of the mystical experience in terms of what exists. A reliable mystical experience and a mystical hallucination would be identical. After all, why do hallucinations deceive us? Because their content is the same as that of a reliable experience. When I see a mirage of an oasis, the water looks as blue and sparkling as real water. Otherwise, I would not be fooled. If the Absolute exists in a realm with content and nothing else, a real Absolute and a hallucinatory Absolute afford exactly the same experience. And there is no price to pay for the hallucinatory Absolute. I cannot rush forward to dip my hand in it and find that it engendered false expectations: it arouses no expectations at all about the "future".

If there is no test to discover whether a mystical experience is reliable or hallucinatory, it is a mistake to classify it as one or the other. Arousing no expectations, such an experience is neither reliable nor deceptive. The pantheist should glory in the view of the universe it offers and say nothing about whether that view reveals a real attribute of the universe. It is simply beside the point: the view is the thing. We have said enough to suggest that the view, the mystical experience itself, is well worth having.

Imagine a world in which everyone is color-blind but in which I suddenly begin to dream in color. In my dream my wife's hair, which has always looked a sort of off-white, is brilliant red. What sense would it make to ask, "Is my wife's hair really red?" It will never appear to be so in waking life. The color content of the dream has no relevance to the colors of everyday life. How can it pose questions about what everyday life is "really like"? Be happy with the view the dream affords: it affords brilliant colors that were hitherto beyond your experience.

If a reliable and a hallucinatory mystical experience are indistinguishable, effectively identical, why care about whether the Absolute exists? We would have the same experience in either event. To reply that if the Absolute did not really exist we would not get the mystical experience is not necessarily the case. None of our hallucinations are caused by independently existing objects, yet we have them anyway. Value the mystical experience for what it is and forget about whether the Absolute exists or not.

Religion arouses strong emotions. In Rudyard Kipling's *Stalky & Co.* a master tells a student that unless he can attest to a belief in a personal god by three o'clock the next afternoon, he will be expelled. A rationalist orator used to begin every speech with the following: "There is no god. There is no god. There is no god at all. He who invented god is a fool. He who propagates god is a scoundrel. He who worships god is a barbarian." This was not Richard Dawkins. It was the sage of the Dravidian Progressive Movement, Erode Venkata Ramasamy (1879–1973).

I am an atheist but not a mad dog atheist. I get annoyed when educated people accept traditional Christianity without reflection, but I respect the two great transcendental experiences, true faith and mysticism. Neither vouches for the existence of anything. When this is realized, true faith is vulnerable because its essence is communication with God and therefore it presupposes God's existence. The mystical experience is not vulnerable because it does not presuppose the existence of the Absolute. The experience itself exhausts whatever significance the existence of the Absolute, either as object or attribute, may have.

CHANGING THE PERSON YOU ARE

Over my lifetime I have been interested in many problems: the theory and measurement of intelligence; whether races and genders have much the same genes for intelligence; the wicked execution of the mentally retarded under some jurisdictions; the moral disaster that is American foreign policy; the benefits of democratic socialism for America and New Zealand; and why universities educate badly. But none of these has meant as much to me as the problems of philosophy. You may wonder, then, why I have never studied or taught in a philosophy department. Perhaps it is because matters of philosophy are so intensely personal. To bare one's soul over tea every day would be an unwelcome experience.

It is hard to believe that any reflective person would want to just absorb the philosophy of his or her time and place. I hope that after reading this book you will want to seize control of the personal philosophy that governs your life, revise it in the light of reason, and recreate yourself as a fully conscious human being.

Above all, I hope you will think deeply about ethics. There is nothing external to the self that can tell you what is right and wrong: you must choose. And you can choose from among all the ideals that human beings have ever cherished.

The ancients believed we should try to establish a society that emphasized many great goods—human happiness, justice, truth, beauty, delight in diversity, and the dignity of man—and reject mores that had proved stultifying, such as materialism, militarism and the cult of popularity.

Subsequent thinkers elevated certain of these goods—human happiness, justice, and tolerance—above all the others in the mistaken belief they could be shown to have objective status. Finally, the good society was reduced to one that satisfied as many human demands as possible, opening the door to the notion of "the market" as the vehicle of the good.

Since humane ideals cannot be shown to have objective status, they live on only thanks to people who are passionately committed to them. Acting on these ideals will put you in the company of people whose devotion to humanity has made them into a band of brothers and sisters striding across time toward the holy city (see box opposite).

This does not mean that adopting ideals is enough. The ideals must be defended in moral debate. That debate should be based on the rules of logic, the teachings of science, and a plausible social dynamic for using your ideals to organize a human society. It is not enough to show that your opponents fail these tests: you must show that you can pass them. This means purging your ideals of sentimental content, and being sufficiently aware of what they mean in practise to answer objections such as the meritocracy thesis and its claim that humane ideals self-destruct.

THE COMPANY TO KEEP
As well as Plato, Aristotle and Jesus:

Hillel (110 BC–10 AD). A great teacher of Judaism, who said: "What is hateful to thee, do not do unto thy fellow man: this is the whole law; the rest is mere commentary." Babylonian Talmud, Shabbat 31a.

Ida Bells Wells (1862–1931). A black woman who in 1884 refused to give up her seat to a white on a train, thus anticipating Rosa Parks (see below). She was a leader in the campaign against lynching.

Emily Wilding Davison (1872–1913). The bravest of those who fought to get British women the vote. She died in 1913 when run over by the king's horse in the Epsom Derby (she was trying to drape a suffragette flag around its neck). On census night, 1911, she had hidden in Westminister so she could list her residence as "House of Commons".

Kier Hardie (1856–1915). Upon the birth of Edward VIII, this great Scottish socialist and Independent Labour MP told the House of Commons: "From his childhood onward this boy will be surrounded by sycophants and flatterers by the score and will be taught to believe himself as of superior creation."

Eugene Victor Debs (1855–1926). Speaking to workers in Detroit in 1906, this labor organizer said: "I would not lead you into the Promised Land if I could, because if I led you in, someone else would lead you out."

George Orwell (1903–1950). The novelist who, when he arrived in Spain on Christmas Day, 1936, simply said: "I have come to fight against fascism."

Erich Maria Remarque (1898–1970). The novelist who so irritated Hitler that the Nazi leader beheaded his sister.

Mahatma Gandhi (1869–1948). The leader of India's nationalist movement who used passive resistance to free his country from British rule.

Martin Luther King (1929–1968). The civil rights leader who used passive resistance to free American blacks from segregation.

Jenny Lee (1904–1980). Perhaps the most principled British parliamentarian of the twentieth century, Lee opposed her husband, Aneurin Bevan, when he supported the United Kingdom's acquisition of nuclear weapons.

Rosa Parks (1913–2005). This black woman refused to give up her seat on a bus to a white man and thereby sparked the Montgomery bus boycott, which fired the American civil rights movement.

You must decide whether or not to praise and blame other people for their behavior. Praise and blame are appropriate only if the present self makes truly free choices in morally significant cases. This is a factual question that may ultimately be resolved by science. However, it is unlikely that science will reach a verdict in the foreseeable future, so we are left with an existential dilemma. For the time being we must either praise and blame, and thus assume people can (at least sometimes) exercise free will, or not praise and blame, and thus assume people's behavior is always determined.

If you are making such a decision, reason provides no guidance. You can be guided only by moral criteria, such as what contributes most to interpersonal relationships—perhaps moral responsibility—and what would do most to humanize the legal code—perhaps no-fault.

I myself changed while writing this book. I had never before fully realized how omnipresent science was in my thinking. Science can never replace philosophy, if only because we have to ponder its implications for what we are and what we believe. Nor can it make a case that it has a monopoly on telling us what exists, or even that it is in accord with reason. But it was science that forced us to set a higher standard of rationality in other areas: we could no longer believe that we were the center of the universe, that we were made by a god, that this life was less important than the next, that our ideals were someone or something else's creation, that we could take free will for granted.

Ever since Galileo, the task of philosophy has been to come to terms with science, and perhaps after four hundred years we are ready to make our peace. Wonderful things await. Soon science may tell us whether we live on the only universe that exists or one out of many. Eventually it may tell us whether our behavior is free or determined. It can still the voices of tribe, race, church and nation. Until the end of human history, if we listen carefully it will tell us when we need new bridging propositions to link our humane proclivities with the latest news about reality. It is a great privilege to be a literate person in a scientific age.

FURTHER READING

Before sampling this list, which includes the books and authors mentioned in the text, you may want to read Bertrand Russell's *History of Western Philosophy*, available in most libraries. Professional philosophers sometimes snipe at Russell but he writes wonderfully and his book will put what you read into the context of time and place.

I have divided into two the fifty books or articles I recommend. Those in **A LIST** are particularly readable. **B LIST** is for those who want to delve more deeply. Many works are classics that come in numerous editions. In most cases the most recent paperback edition has been cited. **E** indicates that an ebook version or versions are available; more may have become available since this book went to press.

A LIST

A. J. Ayer, *Language, Truth and Logic*: Penguin, London, 2001. (First published 1936.) **E.**

Ruth Benedict, *Patterns of Culture*: Mariner Books, Boston, 2006. (First published 1934.)

Alan S. Blinder, *Hard Heads, Soft Hearts: Tough-minded Economics for a Just Society*: Basic Books, New York, 1988.

Richard Dawkins, *The Blind Watchmaker*: Penguin, London, 2006. (First published 1986.) **E.**

Daniel C. Dennett, *Freedom Evolves*: Penguin Books, London, 2004. **E.**

Fyodor Dostoyevsky, *The Brothers Karamazov*: W.W. Norton, New York, 2011. (First published 1880.) **E.**

F. Scott Fitzgerald, *The Great Gatsby*: Penguin, New York, 2007. (First published 1925.) **E.**

James R. Flynn, *How to Defend Humane Ideals: Substitutes for Objectivity*: University of Nebraska Press, Lincoln, Nebraska, 2000.

John Gribbin, *Deep Simplicity: Chaos, Complexity and the Emergence of Life*: Penguin, London, 2005. **E.**

R.M. Hare, *Freedom and Reason*: Oxford University Press, New York, 1977. **E.**

Richard J. Herrnstein and Charles Murray, *The Bell Curve: Intelligence and Class Structure in American Life*: Free Press, New York, 1996. **E.**

Aldous Huxley, *Island*: HarperCollins, New York, 2009. (First published 1962.) **E.**

William James, "The Moral Philosophy and the Moral Life" in *The Will to Believe and Other Essays in Popular Philosophy*: Cosimo Books, New York, 2006. (Essay first published 1897.) **E.**

William James, *Varieties of Religious Experience*: Dover Publications, Mineola, New York, 2003. (First published 1902.) **E**.

Søren Kierkegaard, *Fear and Trembling*: Cambridge University Press, Cambridge, UK, 2006. (First published 1843.) **E**.

Thomas Kuhn, *The Structure of Scientific Revolutions*: University of Chicago Press, Chicago, 2012. (First published 1962.) **E**.

John Stuart Mill, "Nature" in *Nature, the Utility of Religion, and Theism*: Adamant Media Corporation, Boston, 2000. (First published 1874.) Also chapter 4 of *Utilitarianism*: Hackett Publishing, Cambridge, Massachusetts, 2002. (First published 1863.) **E**.

Friedrich Nietzsche, *Beyond Good and Evil*: Penguin, London, 2003.(First published 1886.) **E**.

Steven Pinker, *The Blank Slate: The Modern Denial of Human Nature:* Penguin, London, 2002. **E**.

Plato, *The Republic:* Penguin, London, 2007. (Written circa 380 BC.) **E**.

Thomas W. Pogge, *World Poverty and Human Rights*: Polity, Cambridge, UK, 2007.

Walter T. Stace, *The Teachings of the Mystics*: Mentor Books, Dublin, 1960.

R.H. Tawney, *The Acquisitive Society*: Forgotten Books: Charleston, South Carolina, 2010. (First published 1920.) **E**.

Thorstein Veblen, *The Theory of the Leisure Class*: Oxford University Press, Oxford, 2009. (First published 1899.) **E**.

Ludwig Wittgenstein, "A Lecture on Freedom of the Will" in *Philosophical Investigations*, April 1989.

B LIST

Aristotle, *Nicomachean Ethics*: University of Chicago Press, Chicago, Illinois, 2012. (Written circa 330 BC.) **E**.

Aristotle, *The Politics*: Dover Publications, Mineola, New York, 2000. (Written circa 328 BC.) **E**.

George Berkeley, *Principles of Human Knowledge and Three Dialogues*: Penguin, London, 2004. (*Principles of Human Knowledge* first published 1710; *Three Dialogues between Hylas and Philonous* first published 1713.) **E**.

Cyril Burt, "Class Differences in General Intelligence: III" in *British Journal of Statistical Psychology*, May 1959. Also published online in *British Journal of Mathematical and Statistical Psychology*, August 2011.

Cyril Burt, "Intelligence and Social Mobility" in *British Journal of Statistical Psychology*, May 1961. Also published online in *British Journal of Mathematical and Statistical Psychology*, August 2011.

James R. Flynn, "IQ Trends over Time: Intelligence, Race and Meritocracy" in *Meritocracy And Economic Inequality*, Kenneth Arrow, Samuel Bowles, Steven N. Durlauf (eds): Princeton University Press, Princeton, New Jersey, 2000.

Harry G. Frankfurt, "Alternate Possibilities and Moral Responsibility" in *The Journal of Philosophy*, December 1969.

Thomas Hobbes, *Leviathan*: Penguin, London, 1985. (First published 1651.) **E.**

David Hume, *A Treatise of Human Nature*: Dover Publications, Mineola, New York, 2003. (First published 1961.) **E.**

Immanuel Kant, *Critique of Pure Reason*: Penguin, London, 2007. (First published 1781.) (Do not try to read this without help.) **E.**

John Locke, *An Essay Concerning Human Understanding*: Oxford University Press, Oxford, 2008. (First published 1690.) **E.**

John Locke, *Essays on the Law of Nature and Associated Writings*: Clarendon Press, Oxford, 2002. ("Essays on the Law of Nature" first published 1663–1664.)

Michael Maher, *Psychology: Empirical and Rational*: Out of print; available in some libraries. (First published 1915.) **E.**

G.E. Moore, *Principia Ethica*: Cambridge University Press, Cambridge, UK, 1993. (First published 1903.)

Alan Musgrave, *Essays on Realism and Rationalism*: Editions Rodopi, Amsterdam, 1999.

Kai Nielsen, *Equality and Liberty: A Defense of Radical Egalitarianism*: Rowman & Littlefield, Lanham, Maryland, 1984.

H.J. Paton, *Kant's Metaphysic of Experience: A Commentary on the First Half of the "Kritik der Reinen Vernunft"*: Out of print; available in some libraries. (First published 1936.)

Anton C. Pegis (ed.), *Introduction to St. Thomas Aquinas*: McGraw-Hill, New York, 1965.

Karl Popper, *The Poverty of Historicism*: Routledge, London, 2002. (First published 1957.)

John Rawls, *A Liberal Theory of Justice*: Oxford University Press, Oxford, 1973.

Rush Rhees, "Some Developments in Wittgenstein's View of Ethics" in *Philosophical Review*, 1965.

P.F. Strawson, "Freedom and Resentment" in *Freedom and Resentment and Other Essays*: Routledge, Oxford, 2008. (First published 1974.) **E.**

Tarmo Strenze, "Intelligence and Socioeconomic Success: A Meta-Analytic Review of Longitudinal Research" in *Intelligence*, 2007.

Nicholas L. Sturgeon, "Moral Explanations" in *Essays on Moral Realism*, Geoffrey Sayre-McCord (ed.): Cornell University Press, Ithaca, New York, 1988.

Bas van Fraassen, *The Scientific Image*: Clarendon Press, Oxford, 1980.

TEASERS

Here are a few problems that you need not confront to live your life. If you do not enjoy them, probably you should not become a professional philosopher unless you specialize in ethics, political philosophy or the philosophy of science.

WHAT AM I?

First, a problem of personal identity—that is, the conditions under which a person is said to be identical to himself or herself through time.

Imagine someone has played a trick on you ever since you were born. At birth, thousands of clones of you were made. During every day you live, each is imprinted with the effects of everything you experience. Every night, while you are asleep, you are exterminated and a clone substituted, so the person who wakes up in the morning is identical to what you would have been upon awakening. "You" would never know the difference.

This may have been happening to you throughout what you call "your" life. Now that you know, does it disturb you? A lot of people say yes, because, whatever this succession of creatures is, it is not me.

We now revise the scenario. From birth your parents have known that you suffer from a unique medical condition.

Every twenty-four hours your brain dies at three a.m. The doctors offer a remedy: brain transplants can be done. Even better, an hour before your brain dies all of its characteristics can be transferred to a virgin brain, one that has been wiped clean. Every night, while you were asleep, that was done at two a.m. Does that change how you feel? And if so why?

WHAT ARE OTHER PEOPLE?

The solipsist must be a realist and grant the independent existence of things other than himself, including other people. However, the core of solipsism is whether it is legitimate to assume that other people have minds.

Imagine that some day we create robots that do everything a human being can do, but without having a conscious mental life. You know that you have thoughts and emotions but how do you know that anyone else does? You cannot actually think anyone else's thoughts or feel anyone else's emotions. The fact that the robots have been programmed to talk and look and behave as if they have a conscious life is no proof. The fact they "know" things that you do not—higher mathematics, for example, or Russian language—just shows what good thinking machines they are.

You do not know they are robots of course, but the evidence at hand gives neither their consciousness nor their lack of consciousness an advantage. Either possibility is consistent with the evidence.

Different philosophers have different responses to this. Here's mine. When I teach others something I do not use brain surgery. To be effective, I have to lead them through the same mental steps I had to take when I learned it. Of all the millions of things that might program their thinking machines, it seems odd that the only one that works is duplication of my mental processes. If they are not "conscious" of the concepts I had to entertain to understand Plato's divided line, they must simulate that consciousness very closely.

The same is true of motivation. Of all the things that could trigger their thinking machines into activity, it seems odd that only things such as incentives work. I surprise them with something they do not know but find interesting. It looks as if certain emotions, such as wanting to understand, are necessary to get their thinking going. But this may be cheating. Have I not assumed they are machines that can simulate everything human beings can do, and learning is just one of those things?

Very well, but I should take stock of certain things that I know. I know who my parents are. If they are robots then two robots have had a non-robot child. I understand enough genetics to know that such a radical difference between parents and offspring is possible only if my genes profited from a mutation. Therefore, I need a plausible hypothesis as to what brought it about. Ideally I would have an identical twin, someone whose genes are identical to my own except for a (post-conception) mutation. Even lacking ideal conditions,

I should have a unique gene or genes that can be identified. As knowledge of DNA sequencing grows, and if comparison with parents and siblings reveals nothing unusual, we ought to be able to eliminate all plausible candidates.

It may be said that the situation is like free will: until science advances, both the "others feel" and "others do not feel" possibilities are open. However, in this case no humane person will feel that both options are open. If we were to treat people as if they did not feel—for example, not anaesthetizing them for surgery—and they really did feel, we would inflict great suffering.

ACKNOWLEDGMENTS

In writing this book, I have at times drawn on my earlier published work. Part One includes updated and condensed material from *How to Defend Humane Ideals* (University of Nebraska Press, 2000). The thoughts on nationalism summarise Chapter 7 of *Beyond Patriotism: From Truman to Obama* and appear with the permission of Imprint Academic. The sixth to eighth chapters restate arguments in *Where Have All the Liberals Gone? Race, Class, and Ideals in America* and appear with the permission of Cambridge University Press. I wish to thank all the publishers who have generously granted their permission.

INDEX

The Torchlight List
Around the World in 200 Books

Jim Flynn

One of the best books of the year
— The Sunday Star-Times

A professor for over forty years, Jim Flynn found fewer and fewer of his students had the time or inclination to read for pleasure. However, they were willing to try books he recommended. He was inspired to create this list: books so wonderful to read, and so revealing about times and places, they take the reader beyond day-to-day concerns into a magic realm of knowledge and imagination.

There have been many books about reading. No other has set out such a brilliant road map for discovering history, science, civilization and the human condition. From Arthur Koestler on the universe to Barbara Tuchman on life in the fourteenth century, F. Scott Fitzgerald on American morality, Chimamanda Ngozi Adichie on civil war in Nigeria, and Robert Fisk on Western power plays in the Middle East, this book will inspire you to reread books you love and discover and relish many new ones.

An ambitious yet intimate celebration of the power of literature
— Finlay Macdonald

Flynn's descriptions are like tantalizing movie trailers—short bursts of images that teleport your mind from one side of the Earth to the other
— Psychology Today

AWA PRESS

ISBN 978-0-9582916-9-9
Available from all good bookstores
and online at www.awapress.com